AF389164

Seed Dormancy

Seed Dormancy

Molecular Control of Its Induction and Alleviation

Editor

Angel J. Matilla

MDPI • Basel • Beijing • Wuhan • Barcelona • Belgrade • Manchester • Tokyo • Cluj • Tianjin

Editor
Angel J. Matilla
Department of Functional
Biology, Life Campus,
Faculty of Pharmacy,
University of Santiago de Compostela (USC)
Spain

Editorial Office
MDPI
St. Alban-Anlage 66
4052 Basel, Switzerland

This is a reprint of articles from the Special Issue published online in the open access journal *Plants* (ISSN 2223-7747) (available at: https://www.mdpi.com/journal/plants/special_issues/seed_AR).

For citation purposes, cite each article independently as indicated on the article page online and as indicated below:

LastName, A.A.; LastName, B.B.; LastName, C.C. Article Title. *Journal Name* **Year**, *Article Number*, Page Range.

ISBN 978-3-03943-653-8 (Hbk)
ISBN 978-3-03943-654-5 (PDF)

Cover image courtesy of Raquel Iglesias-Fernández.

Contents

About the Editor

Angel J. Matilla Professor of Plant Physiology since 1983. Universities of Granada and Santiago de Compostela (Spain). Research Gate: Angel Jesus Matilla Author of 122 scientific publications, director of 21 doctoral theses, obtained 10 research projects, referee of several journals and research projects (EEC, Argentina, Spain, Holland).

 *plants*

Editorial

Seed Dormancy: Molecular Control of Its Induction and Alleviation

Angel J. Matilla

Department of Functional Biology, Life Campus, Faculty of Pharmacy, University of Santiago de Compostela (USC), 15782 Santiago de Compostela, Spain; angeljesus.matilla@usc.es

Received: 29 September 2020; Accepted: 16 October 2020; Published: 21 October 2020

Abstract: A set of seed dormancy traits is included in this Special Issue. Thus, DELAY OF GERMINATION1 (DOG1) is reviewed in depth. Binding of DOG1 to Protein Phosphatase 2C ABSCISIC ACID (PP2C ABA) Hypersensitive Germination (AHG1) and heme are independent processes, but both are essential for DOG1's function in vivo. AHG1 and DOG1 constitute a regulatory system for dormancy and germination. DOG1 affects the ABA INSENSITIVE5 (ABI5) expression level. Moreover, reactive oxygen species (ROS) homeostasis is linked with seed after-ripening (AR) process and the oxidation of a portion of seed long-lived (SLL) mRNAs seems to be related to dormancy release. The association of SLL mRNAs to monosomes is required for their transcriptional upregulation at the beginning of germination. Global DNA methylation levels remain stable during dormancy, decreasing when germination occurs. The remarkable intervention of auxin in the life of the seed is increasingly evident year after year. Here, its synergistic cooperation with ABA to promote the dormancy process is extensively reviewed. ABI3 participation in this process is critical. New data on the effect of alternating temperatures (ATs) on dormancy release are contained in this Special Issue. On the one hand, the transcriptome patterns stimulated at ATs comprised ethylene and ROS signaling and metabolism together with ABA degradation. On the other hand, a higher physical dormancy release was observed in *Medicago truncatula* under 35/15 °C than under 25/15 °C, and genome-wide association analysis identified 136 candidate genes related to secondary metabolite synthesis, hormone regulation, and modification of the cell wall. Finally, it is suggested that changes in endogenous γ-aminobutyric acid (GABA) may prevent chestnut germination, and a possible relation with H_2O_2 production is considered.

Keywords: ROS; DOG1; physical dormancy; long-lived mRNA; monosomes; DNA methylation; auxin and ABA; alternating temperatures; GABA

---

1. Introduction

The seed, a key entity in the life cycle of higher plants, allows and ensures its survival by acquiring primary dormancy (PD) during maturation [1]. The DELAY OF GERMINATION1 (DOG1) protein was identified and characterized as a major regulator of seed dormancy [2,3]. PD is the failure of seeds to germinate although environmental conditions are favorable. Interestingly, some PD-related genes are regulated through the epigenetic control of endosperm-specific gene expression [4,5]. Likewise, nondormant seeds can enter secondary dormancy (SD) upon exposure to unfavorable conditions for germination. Lack of light is a key factor involved in the induction of SD. However, it is not yet confirmed whether PD is a requirement to have the ability to acquire SD [6]. Recently, it was demonstrated that SD is induced in both high- and low-dormancy genotypes and that SD is less responsive to after-ripening (AR) and cold stratification than PD [7]. Maternal ABA is the only phytohormone known to induce, regulate, and maintain PD [8], and ABA levels and ABA signaling play pivotal roles in the regulation of PD and germination [9]. Furthermore, the ABA/ Gibberellins

(GAs) balance is key to controlling PD and germination [10,11]. Thus, seeds of ABA-deficient mutants germinate faster than the wild-type ones, and transgenic plants constitutively expressing the ABA biosynthesis genes maintain deep PD [12]. Seed germination processes are under the control of classical phytohormones, reactive oxygen species (ROS) [13], brassinosteroids [14], strigolactones [15], as well as temperature, nitrate, and light [16,17]. Accordingly, PD and germination are strictly regulated by the modulation of suitable phytohormones, transcription factors, and environmental signaling networks [9]. This regulation mechanism is supposed to be highly conserved [18]. Together, PD and germination are two closely linked physiological traits that have great impacts on the adaptation and survival of seed plants. Although the phytohormones involved in these two traits have been largely identified, their mechanisms of interaction with external factors and how dormancy is broken under different conditions are more elusive. In this Special Issue, some aspects of the regulation of seed dormancy and germination are addressed.

2. Seed Dormancy and Delay of Germination-1 (DOG1) Protein

Due to the great repercussion of seed dormancy in the life of the seed, a great deal of research on PD has been developed in the last few decades. One of the reasons, among several others, is the appearance of pre-harvest sprouting (viviparism) in the mother plant when PD is not triggered. Viviparism is an important problem in cereal production because it reduces crop yield and quality. In other words, knowledge of the initiation, maintenance, and loss of PD is key to understanding how the germination process is triggered. The transcriptional and epigenetic control of dormancy, as well as the great advances in proteomics, have clarified a considerable number of PD mechanisms, which are essential to the survival of higher plants. In 2006, DOG1 was identified as a major Quantitative Trait Loci (QTL) for seed dormancy variability among natural *Arabidopsis thaliana* accessions, and *dog1* T-DNA insertional mutants exhibit reduced seed dormancy [19]. The expression of DOG1 is widely regulated and increases during seed maturation. DOG1 protein levels accumulate during the last phase of embryogenesis and correlate with the depth of PD. However, although DOG1 is relatively stable, DOG1 mRNA disappears quickly after seed imbibition. Given its key role in PD, DOG1 has been extensively studied in recent years. Currently, little is known about the precise molecular mechanism underlying the transcriptional regulation of *DOG1*. Carrillo-Barral et al. [20] present here a detailed update on DOG1. Their review focuses on why DOG1 is a key signaling molecule to coordinate seed life and, very specifically, the acquisition and loss of PD. DOG1 enhances ABA signaling through its binding to PP2C ABA Hypersensitive Germination (AHG1). Likewise, DOG1 suppresses the AHG1 action to enhance ABA sensitivity and impose PD. To carry out this suppression, the formation of the DOG1–heme complex is essential. In contrast, *dog1* mutant seeds, which have scarce endogenous ABA and high GA content, exhibit a non-dormancy phenotype. At the physiological level, DOG1 is tightly regulated by a complex array of transformations that include alternative splicing and polyadenylation, histone modifications, and a *cis*-acting antisense non-coding transcript (asDOG1). The activation of *DOG1* expression leading to increased PD requires that bZIP-transcription factor 67 (bZIP67) be bound to the DOG1 promoter. Although *DOG1* is mainly expressed in seeds, other organs are also capable of doing so.

3. ROS and Nucleic Acid Modifications during Seed Dormancy

Plants have to deal with ROS constantly generated in the cell organelles. Except for certain phases of the plant life cycle (e.g., dry viable seeds), the production of ROS is essentially associated with photosynthesis. When an excess of ROS is produced and a threshold exceeded (e.g., under stress conditions), cellular damage may arise and trigger cell death. To a greater or lesser degree, all ROS are markedly reactive. Thus, they are able to oxidize biological molecules, including lipids, DNA, RNAs, and proteins, RNA being more susceptible to oxidative damage than DNA. Interestingly, we now know that ROS is not always detrimental to the cell. This is what sometimes happens, for example, with the singlet oxygen ($^1O_2$). That is, the $^1O_2$ generated in the light-harvesting complex (LHC) of the chloroplast grana core under excessive light energy, or in the photosystem II reaction center (PSII-RC) of the grana margins under low light energy, may act as a highly versatile signal (i.e., chloroplast-to-nucleus retrograde signaling (ChNRS)) triggering beneficial cell responses. To sustain life, an organism must maintain ROS homeostatic levels. This control involves more than 150 genes in *Arabidopsis*. In contrast, ROS has been correlated with a low degree of seed dormancy. When the ROS level reaches a certain threshold, dormancy is alleviated and the subsequent germination can be initiated. Likewise, seed aging takes place, and an intensive degradation of nucleic acids and proteins occurs. Interestingly, it was recently demonstrated that AtPER1, a seed-specific peroxiredoxin, eliminates ROS to suppress ABA catabolism and GA biosynthesis, and thus improves the PD and make the seeds less sensitive to adverse environmental conditions [21]. In this Special Issue, Katsuya-Gaviria et al. [22] review in detail the biological significance of nucleic acid oxidation caused by ROS during PD and germination. This update also refers to the state of the art regarding DNA and RNA methylation in seed biology. Thus, we can see how ROS increases upon after-ripening (AR) and dormancy release. Interestingly, ROS is located close to the radicle apex during imbibition, whereas oxidative species does not have a certain distribution in these dormant seeds. In support of this, several enzymes that participate in ROS homeostasis have been associated with the germination and AR process. ROS oxidizes nucleic acids at different molecular positions affecting their stability. It is known that 8-hydroxyguanosine (8-OHG) is the most habitual oxidative nucleoside in RNA molecules. Likewise, the oxidation of a fraction of seed long-lived (SLL) mRNAs seems to be related to dormancy release and seed aging. However, how the SSL mRNAs involved in germination are preserved from oxidation during dry seed storage is not yet clear. Recently, it was proven that ~17% of the SLL mRNAs are specifically associated with monosomes and are translationally upregulated during seed germination. For this, the formation of the SSL mRNA–RNA binding protein (RBP)–monosome complex seems to be key in the process of safeguarding these SSL mRNAs. Together, the association with monosomes likely protects the SSL mRNAs needed during early seed imbibition in a state ready for translation. Their review clearly specifies all of the above. Furthermore, DNA methylation is a well-known epigenetic mechanism of controlling gene expression. During *A. thaliana* embryogenesis, there is a global increase in CHH-context methylation. Global DNA methylation levels remain stable during seed dormancy, decreasing when germination occurs. Up to now, the presence of specific DNA methylation markers associated with dormancy or germination transcriptomes remains to be elucidated.

4. Auxins and Seed Dormancy

In addition to higher plants, auxin is a signaling molecule that is present in living organisms such as algae, moss, liverworts, lycophytes, and microorganisms. Auxin is involved in multi-functional processes during plant growth and development. Recently, auxin signaling was thoroughly reviewed [23]. Regarding the seed, it is now widely accepted that auxin biosynthesis is required for an array of seed developmental processes (e.g., embryogenesis and endosperm development, among others). Current studies have elucidated that auxin is also involved in the transition from PD to germination. Recent studies have shown that auxin possesses positive effects on seed dormancy, being (in conjunction with ABA) the second known hormone that induces seed dormancy. Thus, it was demonstrated, for the first time at the molecular level, a role for auxin in PD through stimulation of ABA signaling, identifying auxin as a dormancy promotor [24]. Matilla (2020) [25] carries out here an in-depth review of the participation of auxin in embryogenesis, PD, and germination. The dynamic of expression and localization (i.e., proembryo, hypophysis, and suspensor) of several key genes for the biosynthesis and transport of auxins in the globular phase was carefully checked (Figure 1 of the review). The bHLH49 transcription factor appears to be a notable mediator of the auxin-dependent suppression of embryo identity in suspensor cells. Likewise, synthesis, transport, and compartmentalization of auxins are crucial for the ovule, endosperm, and seed-coat (SC) development. Auxin transport from the endosperm to the integuments is regulated by AGAMOUS-LIKE 62 (AGL62), the encoding gene of which is specifically expressed in the endosperm to suppress its cellularization. In the absence of

AGL62 (i.e., agl62 mutants), auxin remains trapped in the endosperm and the SC fails to develop (i.e., seed abortion). The application of auxin represses soybean seed germination through decreasing the ABA/GA ratio. Jointly, it is suggested that auxin acts synergistically with ABA to promote PD and inhibit germination. Recent biochemical and genetic evidence supports the involvement of auxins in PD. In this process, the participation of the transcriptional regulator ABA INSENSITIVE3 (ABI3) is critical, revealing a cross-talk between auxin and ABA signaling. Recent information demonstrates that auxin acts downstream of ABA to promote seed germination. However, it is still unknown if any process exists in which ABA acts downstream of auxin. An exhaustive analysis of the auxin responsiveness of ABA biosynthesis, transport, and signaling mutants will be required for this.

5. Gene Expression Patterns and Physiological Response Associated with Release of Dormancy under Alternating Temperatures

Although alternating temperatures (ATs) are more effective than constant ones in stimulating germination of some seeds, little is known at the physiological and molecular level about the regulation of the breaking of dormancy by ATs. It now seems clear that the convergence between ROS signaling and classical phytohormones participates in this dormancy-breaking process. Huarte et al. [26] now turn their attention to this breaking mechanism using after-ripening cardoon seeds as a biological system. Previously, it was proven that fluctuating temperatures terminate dormancy in this seed by turning off ABA synthesis and reducing ABA signaling, but not stimulating GA synthesis or signaling [27]. In this work, an advance in the break dormancy knowledge is carried out through large-scale gene expression. The transcriptome patterns stimulated at ATs comprised ethylene and ROS signaling and metabolism together with ABA degradation. In parallel, the upregulation of ethylene metabolism under AT conditions is also supported by physiological analysis. Interestingly, ROS depletion hampers the breakage process.

6. Effects of GABA on the Germination of Recalcitrant Seeds: Implications on Primary Dormancy

γ-Aminobutyric acid (GABA), a non-protein amino acid, is an important component of the free amino acid pool of living organisms. The enzymes involved in its metabolism are evolutionarily very conserved. Recently, the GABA implications in plant growth and development have been updated [28]. Thus, genetic and physiological studies have proven that GABA is involved in barley aleurone metabolism. Furthermore, the scant current evidence on the mechanism by which GABA acts as a signaling molecule in plants has been also reviewed [29]. The recent identification of a GABA receptor indicates that GABA is a signaling molecule and not just a metabolite [30]. Vigabatrin is a specific GABA transaminase inhibitor that inhibits GABA degradation. In this Special Issue, Du et al. [31] now report that high GABA levels exist in the chestnut recalcitrant viable seeds before germination. Likewise, they also suggest that endogenous GABA may play a specific role in the germination. Exogenous GABA and vigabatrin induced an accumulation of H_2O_2, possibly contributing to the inhibition of chestnut seed germination. In parallel, the authors point out that this inhibition may be due to an alteration in the balance between carbon and nitrogen metabolism, especially the free amino acid contents before germination. Together, the results presented here suggest that changes in GABA levels in chestnut seeds may prevent seed germination.

7. Physical Dormancy Release in *Medicago truncatula* Seeds is Related to Environmental Variations

Physical dormancy is caused by water-impermeable palisade cells in the SC. It is frequent in legumes, and the factors that release this type of dormancy are hardly known [32]. Temperature and soil moisture oscillations are the major players under natural conditions. Renzi et al. [33] present a study on temperature-related physical dormancy release in seeds of *Medicago truncatula*. These seeds exhibit both physical and physiological dormancy, the latter being non-deep. Seed dormancy release varied among accessions and years, and this could potentially act as a mechanism that favors the persistence of the seed in the soil and helps to distribute genetic diversity through time. However, comparing the results obtained with others recently published, the authors suggest that dormancy is an adaptation securing population survival in less predictable conditions. Moreover, unpredictable natural environments can select earlier within-season germination phenology. On the contrary, although dormancy is genetically determined, it also depends on the environmental conditions experienced by the mother plant and the subsequent status of the seed. In this work, the authors carry out a detailed and in-depth discussion based on the results obtained and those previously published on the genetic basis of the release of seed dormancy in legumes.

Funding: This research received no external funding.

Acknowledgments: I especially wish to thank Sylvia Guo. Without their important collaboration, this Special Issue would not have come to fruition.

Conflicts of Interest: The author declares no conflict of interest.

References

1. Chahtane, H.; Kim, W.; López-Molina, L. Primary seed dormancy: A temporally multilayered riddle waiting to be unlocked. *J. Exp. Bot.* **2017**, *68*, 857–869. [CrossRef] [PubMed]
2. Née, G.; Xiang, Y.; Soppe, W.J.J. The release of dormancy, a wake-up call for seeds to germinate. *Curr. Opin. Plant Biol.* **2017**, *35*, 8–14. [CrossRef] [PubMed]
3. Footitt, S.; Walley, P.G.; Lynn, J.R.; Hambidge, A.J.; Penfield, S.; Finch-Savage, W.E. Trait analysis reveals DOG1 determines initial depth of seed dormancy, but not changes during dormancy cycling that result in seedling emergence timing. *New Phytol.* **2019**, *225*, 2035–2047. [CrossRef] [PubMed]
4. Dekkers, B.J.W.; Pearce, S.P.; van Bolderen-Veldkamp, R.P.M.; Holdsworth, M.J.; Bentsink, L. Dormant and after-ripened Arabidopsis thaliana seeds are distinguished by early transcriptional differences in the imbibed state. *Front. Plant Sci.* **2016**, *7*, 1323. [CrossRef]
5. Iwasaki, M.; Hyvärinen, L.; Piskurewicz, U.; López-Molina, L. Non-canonical RNA-directed DNA methylation participates in maternal and environmental control of seed dormancy. *eLife* **2019**, *8*, e37434. [CrossRef]
6. Soltani, E.; Baskin, J.M.; Baskin, C.C. A review of the relationship between primary and secondary dormancy, with reference to the volunteer crop weed oilseed rape. *Weed Res.* **2019**, *59*, 5–14. [CrossRef]
7. Buijs, G.; Vogelzang, A.; Nijveen, H.; Bentsink, L. Dormancy cycling: Translation-related transcripts are the main difference between dormant and non-dormant seeds in the field. *Plant J.* **2020**, *102*, 327–339. [CrossRef]
8. Penfield, S. Seed dormancy and germination. *Curr. Biol.* **2017**, *27*, R853–R909. [CrossRef] [PubMed]
9. Mine, A. Interactions between ABA and other hormones. In *Abscisic Acid in Plants*; Seo, M., Marion-Poll, A., Eds.; Academic Press: Cambridge, MA, USA, 2019; ISBN 9780081026205.
10. Tuan, P.A.; Kumar, R.; Rehal, P.K.; Toora, P.K.; Ayele, B.T. Molecular mechanisms underlying abscisic acid/gibberellin balance in the control of seed dormancy and germination in cereals. *Front. Plant Sci.* **2018**, *9*, 668. [CrossRef]
11. Vishal, B.; Kumar, P.P. Regulation of seed germination and abiotic stresses by gibberellins and abscisic acid. *Front. Plant Sci.* **2018**, *9*, 838. [CrossRef]
12. Liao, Y.; Bai, Q.; Xu, P.; Wu, T.; Guo, D.; Peng, Y.; Zhan, H.; Deng, X.; Chen, X.; Luo, M.; et al. Mutation in rice *Abscisic Acid2* results in cell death, enhanced disease resistance, altered seed dormancy and development. *Front. Plant Sci.* **2018**, *9*, 405. [CrossRef] [PubMed]
13. Bailly, C. The signalling role of ROS in the regulation of seed germination and dormancy. *Biochem. J.* **2019**, *476*, 3019–3032. [PubMed]
14. Kim, S.Y.; Warpehac, K.M.; Huber, S.C. The brassinosteroid receptor kinase, BRI1, plays a role in seed germinationand the release of dormancy by cold stratification. *J. Plant Physiol.* **2019**, *241*, 153031. [CrossRef]
15. Zwanenburg, B.; Blanco-Ania, D. Strigolactones: New plant hormones in the spotlight. *J. Exp. Bot.* **2018**, *69*, 2205–2218.
16. Duermeyer, L.; Khodapanahi, E.; Yan, D.; Krapp, A.; Rothstein, S.J.; Nambara, E. Regulation of seed dormancy and germination by nitrate. *Seed Sci. Res.* **2018**, *23*, 2003–2010.
17. Yan, A.; Chen, Z. The control of seed dormancy and germination by temperature, light and nitrate. *Bot. Rev.* **2020**. [CrossRef]

18. Carbonero, P.; Iglesias-Fernández, R.; Vicente-Carbajosa, J. The AFL subfamily of B3 transcription factors: Evolution and function in angiosperm seeds. *J. Exp. Bot.* **2017**, *68*, 871–880. [CrossRef]

19. Bentsink, L.; Jowett, J.; Hanhart, C.J.; Koornneef, M. Cloning of *DOG1*, a quantitative trait locus controlling seed dormancy in *Arabidopsis*. *Proc. Natl. Acad. Sci. USA* **2006**, *103*, 17042–17047. [CrossRef]

20. Carrillo-Barral, N.; Rodríguez-Gacio, M.C.; Matilla, A.J. Delay of germination-1 (DOG1): A key to understanding seed dormancy. *Plants* **2020**, *9*, 480. [CrossRef]

21. Chen, H.; Ruan, J.; Chu, P.; Fu, W.; Liang, Z.; Tong, J.; Xiao, L.; Liu, J.; Li, C.; Huang, S. AtPER1 enhances primary seed dormancy and reduces seed germination by suppressing the ABA catabolism and GA biosynthesis in *Arabidopsis* seeds. *Plant J.* **2020**, *101*, 310–323. [CrossRef]

22. Katsuya-Gaviria, K.; Caro, E.; Carrillo-Barral, N.; Iglesias-Fernández, R. Reactive oxygen species (ROS) and nucleic acid modifications during seed dormancy. *Plants* **2020**, *9*, 679. [CrossRef] [PubMed]

23. Leyser, O. Auxin signaling. *Plant Physiol.* **2018**, *176*, 465–479. [CrossRef] [PubMed]

24. Liu, X.; Zhang, H.; Zhaob, Y.; Fenga, Z.; Lia, Q.; Yangc, H.Q.; Luand, S.; Lie, J.; He, Z.H. Auxin controls seed dormancy through stimulation of abscisic acid signaling by inducing ARF-mediated ABI3 activation in Arabidopsis. *Proc. Natl. Acad. Sci. USA* **2013**, *110*, 15485–15490. [CrossRef] [PubMed]

25. Matilla, A.J. Auxin: Hormonal signal required for seed development and dormancy. *Plants* **2020**, *9*, 705. [CrossRef]

26. Huarte, H.R.; Puglia, G.D.; Prjibelski, A.D.; Raccuia, S.A. Seed transcriptome annotation reveals enhanced expression of genes related to ROS homeostasis and ethylene metabolism at alternating temperatures in wild cardoon. *Plants* **2020**, *9*, 1225. [CrossRef] [PubMed]

27. Huarte, H.R.; Luna, V.; Pagano, E.A.; Zavala, J.A.; Benech-Arnold, R.L. Fluctuating temperatures terminate dormancy in *Cynara cardunculus* seeds by turning off ABA synthesis and reducing ABA signaling, but not stimulating GA synthesis or signaling. *Seed Sci. Res.* **2014**, *24*, 79–89. [CrossRef]

28. Carillo, P. GABA shunt in durum wheat. *Front. Plant Sci.* **2018**, *9*, 100. [CrossRef]

29. Fromm, H. GABA signaling in plants: Targeting the missing pieces of the puzzle. *J. Exp. Bot.* **2020**. [CrossRef]

30. Bown, A.W.; Shelp, B.J. Plant GABA: Not just a metabolite. *Trend Plant Sci.* **2016**, *21*, 811–813. [CrossRef]

31. Du, C.; Chen, W.; Wu, Y.; Wang, G.; Wu, Y.; Zhao, J.; Sun, J.; Ji, J.; Yan, D.; Jiang, Z.; et al. Effects of GABA and vigabatrin on the germination of Chinese chestnut recalcitrant seeds and its implicacions for seed dormancy and storage. *Plants* **2020**, *9*, 449. [CrossRef]

32. Rubio de Casas, R.; Willis, C.G.; Pearse, W.D.; Baskin, C.C.; Baskin, J.M.; Cavender-Bares, J. Global biogeography of seed dormancy is determined by seasonality and seed size: A case study in the legumes. *New Phytol.* **2017**, *214*, 1527–1536. [CrossRef] [PubMed]

33. Renzi, J.P.; Duchoslav, M.; Brus, J.; Hradilova, I.; Pechanec, V.; Václavek, T.; Machakova, J.; Hron, K.; Verdier, J.; Smýkal, P. Physical dormancy release in *Medicago truncatula* seeds is related to environmental variations. *Plants* **2020**, *9*, 503. [CrossRef] [PubMed]

Publisher's Note: MDPI stays neutral with regard to jurisdictional claims in published maps and institutional affiliations.

Article

Seed Transcriptome Annotation Reveals Enhanced Expression of Genes Related to ROS Homeostasis and Ethylene Metabolism at Alternating Temperatures in Wild Cardoon

Hector R. Huarte [1,†], Giuseppe. D. Puglia [2,*,†], Andrey D. Prjibelski [3] and Salvatore A. Raccuia [2]

[1]   CONICET/Faculty of Agricultural Sciences, National University of Lomas de Zamora, 1836 Llavallol, Argentina; hrhuarte@gmail.com
[2]   Institute for Agricultural and Forestry Systems in the Mediterranean (ISAFoM), Department of Biology, Agriculture and Food Science (DiSBA), National Research Council (CNR), Via Empedocle, 58, 95128 Catania, Italy; salvatoreantonino.raccuia@cnr.it
[3]   Center for Algorithmic Biotechnology, Institute of Translational Biomedicine, St. Petersburg State University, 199004 St. Petersburg, Russia; andrewprzh@gmail.com
*   Correspondence: giuseppediego.puglia@cnr.it; Tel.: +39-0956139914
†   These authors equally contributed to this work and must be considered both as first author.

Received: 7 August 2020; Accepted: 13 September 2020; Published: 18 September 2020

Abstract: The association among environmental cues, ethylene response, ABA signaling, and reactive oxygen species (ROS) homeostasis in the process of seed dormancy release is nowadays well-established in many species. Alternating temperatures are recognized as one of the main environmental signals determining dormancy release, but their underlying mechanisms are scarcely known. Dry after-ripened wild cardoon achenes germinated poorly at a constant temperature of 20, 15, or 10 °C, whereas germination was stimulated by 80% at alternating temperatures of 20/10 °C. Using an RNA-Seq approach, we identified 23,640 and annotated 14,078 gene transcripts expressed in dry achenes and achenes exposed to constant or alternating temperatures. Transcriptional patterns identified in dry condition included seed reserve and response to dehydration stress genes (i.e., *HSPs*, peroxidases, and *LEAs*). At a constant temperature, we observed an upregulation of ABA biosynthesis genes (i.e., *NCED9*), ABA-responsive genes (i.e., *ABI5* and *TAP*), as well as other genes previously related to physiological dormancy and inhibition of germination. However, the alternating temperatures were associated with the upregulation of ethylene metabolism (i.e., *ACO1, 4,* and *ACS10*) and signaling (i.e., *EXPs*) genes and ROS homeostasis regulators genes (i.e., *RBOH* and *CAT*). Accordingly, the ethylene production was twice as high at alternating than at constant temperatures. The presence in the germination medium of ethylene or ROS synthesis and signaling inhibitors reduced significantly, but not completely, germination at 20/10 °C. Conversely, the presence of methyl viologen and salicylhydroxamic acid (SHAM), a peroxidase inhibitor, partially increased germination at constant temperature. Taken together, the present study provides the first insights into the gene expression patterns and physiological response associated with dormancy release at alternating temperatures in wild cardoon (*Cynara cardunculus* var. *sylvestris*).

Keywords: RNA-Seq; dormancy termination; gene expression; antioxidants; ethylene signaling; environmental signals

---

1. Introduction

Seed dormancy is a continuum process through which dispersed seeds continually sense their surrounding environment perceiving essential information about the most suitable moment to germinate [1,2]. This perception allows modulating seed dormancy level in a cycling way from a high to a low level and vice versa until the suitable germination conditions are met [3]. Environmental temperature, namely constant temperature, acts as a dormancy-alleviation factor, gradually reducing the level of dormancy of the seed population [4]. As the dormancy level is reduced, the ranges of water potential and thermal conditions suitable for germination completion become wider. However, a lot of species still require the presence of some external signals to definitively terminate the dormancy state. Among these, alternating temperatures and light act as dormancy-termination factors removing the ultimate constraint for germination completion once dormancy is sufficiently low [3,4]. Their effect consists of a rapid increase of germination of seeds that have a lowered dormancy degree [5,6]. The daily alternation between low night and high day temperature is an important environmental signal that seeds of some species are adapted to perceive [1,7]. This can provide information on the presence of other plant competitors and the depth of the soil similarly to light [3,8]. This sensing can be very useful, especially for weeds living in variable environments such as the Mediterranean basin [9,10]. Despite the importance of alternating temperatures as a dormancy-termination factor for the completion of germination, little is known about the regulation at the physiological and molecular level of this essential step [11]. Alternating temperatures have recently been found to inhibit abscisic acid (ABA) synthesis through the downregulation of *9-CIS-EPOXYCAROTENOID DIOXYGENASE (NCED)*, an enzyme committed to ABA biosynthesis altering the ABA/GA hormone balance [12]. Otherwise, alternating temperatures may act on decreasing ABA sensitivity, as recently postulated for *Polygonum aviculare* [13]. Beyond the GA/ABA hormone ratio, ethylene is actively involved in the promotion of seed germination and acts antagonistically to ABA during *Arabidopsis thaliana* seed development and several other species [14–17]. Its role in breaking seed dormancy is still not completely ascertained, but there is evidence suggesting that ethylene minimally contributes during dormancy inception, while its major action is during seed imbibition to terminate dormancy and/or initiate germination via crosstalk between ABA and GA pathways [14,18]. This was proposed to determine a decreasing sensitivity to endogenous ABA in concert with GAs to promote these transitional changes leading to germination completion [17]. However, the real magnitude of ethylene contribution to dormancy termination remains to be unveiled. Moreover, there is no evidence of ethylene participation as a part of physiological mechanisms underlying seed exposure to alternating temperatures. Despite reactive oxygen species (ROS) having been considered for a long time as only damaging compounds, in the last decades, they have emerged as key players in seed physiology [19,20]. Recent studies suggest that ROS act as a convergence point of hormonal networks driving cell functioning towards germination through a cross-talk with the major hormonal regulators, i.e., ABA, GA, or ethylene, determining a "ROS wave" [21]. This is carried out by various forms of ROS signaling compounds (e.g., superoxide, hydrogen peroxide and hydroxyl radical) in seeds [19]. In *A. thaliana* the addition in the germination medium of methyl viologen, a ROS-generating compound, partially released seed dormancy, while in sunflower, it alleviated significantly dormancy activating downstream elements of the ethylene signaling pathway but without altering ABA production [20,22,23]. In wild cardoon, previous studies showed an increment of germination in the presence of H_2O_2 [24,25]. On the other hand, when ROS level exceeded a certain value, the activation of antioxidant systems was observed in many species, which allows maintaining ROS homeostasis within the oxidative window for germination [26]. To date, transcriptome investigation has been scarcely applied in seed physiology since it was considered to provide only a partial understanding of the cellular events regulating seed dormancy alleviation or termination processes [27,28]. However, many recent contributions showed that, especially for species for which there is a lack of molecular data, transcripts composition analysis provides new insights on gene interactions and their regulatory mechanisms [29–31]. Microarray analysis showed that thermal oscillations elicited almost immediate large transcriptome changes in leafy spurge seeds exposed to alternating temperatures [5,32]. Moreover, a mitochondrial matrix-localized heat shock protein, HSP24.7, was shown to represent a critical factor that positively controls seed germination via temperature-dependent ROS generation in cottonseed [33]. However, the molecular dynamics during the dormancy termination step remains largely unknown, especially for non-model organisms that lack genetic and physiological data. The botanical species *Cynara cardunculus* L. includes globe artichoke (subsp. *scolymus* (L.) Hegi), cultivated cardoon (var. *altilis* DC.) also known as industrial cardoon for its bioenergy crop uses [34,35] and the wild cardoon (var. *sylvestris* (Lamk) Fiori) that is considered the progenitor of the globe artichoke [36]. Previous studies investigated the germination physiology of the wild variety, demonstrating that alternating temperatures are useful to terminate achenes dormancy causing an abrupt increase of germination percentage, especially in dry after-ripened achenes [24]. This effect was postulated to be triggered by embryo growth potential with a hormonal regulation through a reduction of gibberellins (GAs) and abscisic acid (ABA) ratio and a decrease in ABA sensitivity [12,37]. To date, limited information has been revealed about the transcriptional regulation in cardoon. The recent publication of a low coverage artichoke genome [38], as well as the investigation of cultivated cardoon flowering transcriptome [39], represent novel essential tools to get further insights about the physiological basis of environmental sensing in wild cardoon. In the present study, we analyzed the transcriptome patterns changes associated with imbibition at alternating temperature using after-ripened achenes with a lowered dormancy level to specifically investigate the dormancy termination process. Since transcriptome dynamics associated with the stimulatory effect on dormancy termination of alternating temperature have not been elucidated for any species, we used a gene co-expression approach analysis to identify gene expression associations that may be involved in the regulation of this process. Moreover, to widen our understanding of the underlying physiological modulation, we investigated the changes in ethylene metabolism in achenes exposed to alternating temperatures and the germination response to compounds able to reduce or increase the ROS content. Altogether our results provide further insights to the dormancy termination process stimulated by alternating temperatures including specific transcriptional patterns and regulation of ROS and ethylene levels.

2. Materials and Methods

2.1. Achenes Collection

Mature achenes of wild cardoon, *Cynara cardunculus* var. *sylvestris*, were collected from 20 randomly selected plants exhibiting mature capitula (with a fully expanded pappus and easily detachable achenes from the receptacle) growing at a plot in Llavallol, Buenos Aires Province, Argentina (34°27′ S; 58°26′ W) during January of 2019. After collection achenes from different plants were put together and treated as one lot, cleaned and exposed to dehumidification airflow by 24 h to reach a moisture content of approximately 4–5% on a fresh weight basis assessed using humidity measuring instrument (Rotronic, Ettlingen, Germany). The cleaned achenes lot was stored for 7 months at −18 °C (using a freezer) in tightly closed jars filled to 50% with silica gel to maintain the initial moisture level and silica gel was replaced as soon we observed color turning. This was performed to preserve the physiological state of achenes and to prevent ageing. To alleviate achene dormancy and to analyze the effect of different temperature regimes on dormancy termination, for all the tests performed in the present study, we used dry after-ripened achenes at 35 °C for 21 days as reported in greater detail in Huarte et al. [24].

2.2. General Procedures for Germination Tests

Dry after-ripened achenes were placed in 9-cm diameter Petri dishes over two pieces of filter paper wetted with 7 mL of distilled water or the corresponding treatment solution. Germination tests (four technical replicates of 25 achenes each) were performed in darkness through wrapping in a double layer of aluminum foil each dish. Darkness was used to prevent the interference of light presence as dormancy termination cue. Achenes were imbibed at 20/10 °C (hereafter referred to as alternating temperatures) with a 12 h thermo-period, or 15 °C (constant temperature) in germination chambers with controlled temperature conditions (±1 °C). Moreover, we also used constant temperatures of 10 and 20 °C to observe wild cardoon germination behavior at the minimum and maximum temperatures of the selected alternating thermal regime. Germination was scored daily, and we keep on monitoring it for 14 days after the last achene germination (unless otherwise stated). Achenes with visible radicle protrusion were considered as germinated and then removed. Data were subjected to ANOVA and means were separated using Tukey's test at P, 0.05. Data were analyzed using Infostat.

2.3. Achenes Treatments and RNA Extraction

To annotate the wild cardoon seed transcriptome and analyze the expression dynamics of selected genes related to seed dormancy termination, we exposed achenes to three different conditions: dry achenes, 48 h imbibed achenes at alternating temperature, 48 h imbibed achenes at a constant temperature following the conditions described in the previous section. Each condition was made up of three biological replicas, and for each replica, we used 30 achenes which were immediately immersed in liquid nitrogen and ground as a whole to a fine powder. Total RNA extraction was carried out starting from about 100 mg of the obtained fine powder using RNAeasy Plant Mini Kit (Qiagen, Hilden), with DNase treatment following the manufacturer's protocol. RNA quality and quantity were determined using Eppendorf BioSpectrometer (RNA program) and QIAxcel RNA QC Kit (Qiagen, Hilden) selecting only RNA samples with a RIN/RIS/RQN > 7 to be used for downstream analyses, i.e., RNA-Seq and qRT-PCR.

2.4. Transcriptome Sequencing, Assembly, and Annotation

In the present study, we carried out an explorative transcriptome analysis aimed at the annotation and identification of relevant transcripts in the seed dormancy termination process in wild cardoon. We used two biological replicates for each treatment condition for RNA-Seq analysis. Library preparation and sequencing were outsourced (Eurofins GmbH, Ebersberg, Germany). For each sample, approximately 1 µg of total RNA was used for library preparation applying a strand-specific cDNA libraries synthesis kit (New England Biolabs, Ipswich, MA, USA). The mRNA was selected with a polyA capturing method, fragmented, ligated with adapters, and amplified. Samples from each library were pooled equimolar and paired-end (PE) sequenced using HiSeq2500 (Illumina Technologies, San Diego, CA, USA) platform with chemistry v4 applying the high-output run mode. Illumina reads were analyzed with the FastQC program, and then quality and adaptors, barcodes, polyA and polyT ends were trimmed using Cutadapt v1.16 with default parameters for paired-end reads and Trimmomatic v0.33 [40] in paired-end mode, setting the minimum length to 50bp. Reads were mapped to v.2 of *C. cardunculus* genome available at (www.artichokegenome.unito.it) with Hisat2 aligner [41]. Gene expression levels were estimated with featureCounts [42] using recently updated cardoon gene annotation [39] and expressed transcripts were carried out for further analysis (FPKM > 2). To functionally annotate the obtained transcripts, we aligned them to the publicly available protein databases including NCBI non-redundant (nr) protein database (downloaded in December 2019), using a local BLASTX analysis with an E value cut-off of 10^{-25} and using InterProScan to infer protein function. The results were used with Blast2Go suite program [43] using default parameters to retrieve Gene Ontology (GO) terms and enzyme codes to visualize specific pathways loaded from Kyoto Encyclopedia of Genes and Genomes (KEGG). The composition of genes was investigated through an enrichment analysis of transcriptome using the Fisher's Exact test and False Discovery Rate (FDR) considering the transcriptome analyzed in this study as "test-set" and the annotated transcriptome including several phenological stages of *C. cardunculus* obtained in Puglia et al. [39] as "reference-set". The enriched GO list was, then, analyzed with the AgriGO web application, with Benjamini-Hochberg correction (p-value ≤ 0.01) to limit the representation to the most enriched terms. Moreover, to provide a general overview of the contribution of TFs within the seed dormancy termination process, we searched for sequence homologous in the v4.0 Plant Transcription Factor Database (www.planttfdb.cbi.pku.edu.cn; downloaded in December 2019) using local BLASTX (E value cut-off of 10^{-25}) and we compared their composition among the treatment conditions.

2.5. Differential Gene Expression and Co-Expression Network Analysis

To quantify wild cardoon transcripts expressions, we aligned pre-processed quality-trimmed reads on the reference genome, and we calculated the expression values with the aligned read counts for each transcript. HiSat2 software [41] was used to align the reads on the transcript sequences and HtSeq count [44] was used to evaluate gene expression, in terms of Transcripts per Millions (TPM), from the aligned results. The analysis of differentially expressed genes (DEGs) was carried out with edgeR R package following manual directions for testing multiple conditions. In each analysis, a criterion of |log2(Ratio)| ≥ 2 and an FDR of ≤ 0.01 was used. We run a co-expression analysis on the subset of genes previously identified as differentially expressed using the coseq R package [45] with the K-means approach. The correlation matrix was visualized and analyzed by Cytoscape (version 3.7.2) for co-expression network of genes (http://www.cytoscape.org). To evaluate the transcriptional dynamics of ROS and ethylene pathways across the tested conditions, we selected from DEGs the transcripts with 'antioxidant activity', 'cellular response to stimulus', 'response to endogenous stimulus', 'response to stress', 'seed development' and 'signal transduction' GO terms and we plotted their relative expression as a heatmap. Among this set of transcripts, we selected six genes to be used for real-time PCR analysis for RNA-Seq data validation. For each gene, we differentiated the specific isoform by aligning homologous sequences of *A. thaliana* and wild cardoon using Clustal Omega web server (https://www.ebi.ac.uk/Tools/msa/clustalo/). Identified sequences were used to design specific qRT-PCR primers (Table S1) while as a housekeeping gene we used the actin gene primers already identified in a previous study [39]. Starting from the total RNA extractions, we prepared cDNA libraries using the QuantiTect Kit (Qiagen, Hilden, Germany) and performed real-time PCR reactions on a Rotorgene 6000 cycler (Qiagen, Hilden, Germany) with the QuantiNova SYBR Green Kit (Qiagen, Hilden, Germany). For each treatment condition, we used three biological replicates and three technical replicates of each biological replicates. The fold change in all tissues for each gene was calculated concerning dry achenes condition using the $2^{-\Delta\Delta CT}$ method. The selected genes set were used to validate the expression profiles of RNA-Seq data through a correlation analysis between their expression profiles measured by qRT-PCR and RNA-Seq was calculated with R software.

2.6. Analysis of Ethylene and ROS Regulation at Alternating Temperatures

The effect of ethylene synthesis inhibition on dormancy termination was analyzed by incubating achenes at alternating temperatures in the presence of aminoisobutyric acid (AIB) at doses of 0, 100, 200 and 300 μM (otherwise stated, all chemical compounds were purchased from AG Research, Sigma Argentina) or cobalt chloride ($CoCl_2$) (Anedra, Argentina) at doses of 0, 1.25, 2.5 mM. Similarly, the effect of the inhibition of ethylene signaling on dormancy termination was tested incubating achenes at alternating temperatures the presence of 1-methylcyclopropene (1-MCP) (Smartfresh, Argentina) (0, 25, 50 and 100 μM) and $AgNO_3$ (Anedra, Argentina) (0, 0.25, 0.5 and 1 mM). On the other hand, to evaluate if the presence of an ethylene releasing compound may increase germination at either fluctuating or constant temperature, achenes were imbibed with 2-chloroethyl-phosphonic acid, i.e., ethephon (Tifon, Gleba SA, La Plata, Argentina) at concentrations of 0, 25, 50 and 100 mM. Moreover, the involvement of ROS compounds in seed dormancy termination was investigated incubating achenes at alternating temperatures in the presence of antioxidants, ROS scavengers, ROS donors and ROS synthesis enzyme inhibitors. We used ascorbic acid and glutathione (GSH) at doses of 0, 10, 20, 40, and 60 mM as antioxidant compounds. While, as a ROS scavenger we used Dimethylthiourea (DMTU) for H_2O_2 at a dose of 10 mM. To evaluate NAD(P)H oxidase inhibition (an enzyme related to ROS synthesis) was used Diphenyleneiodonium chloride (DPI) at a dose of 0.1 mM. On the other hand, the effect of methyl viologen (a ROS donor) (0, 0.125, 0.25, 0.5, and 1 mM by 4 h) and Salicylhydroxamic acid (SHAM), a Peroxidase inhibitor (Leymarie et al., 2012) at doses of 0, 2.5, and 5 mM was studied. The effect of SHAM presence was tested at 15 and 20/10 °C.

2.7. Ethylene Measurements

To quantify the different ethylene content produced regarding imbibition temperatures, we imbibed dry after-ripened achenes in water (0.6 mL) on two sheets of filter paper inside vial tubes and incubated for 3, 4 or 5 days at alternating or constant temperatures until ethylene measurement was carried out. Vials were sealed with a septum (natural rubber) and then with parafilm to avoid loss of ethylene. Furthermore, we replicated these imbibition conditions with other achenes from the same batch to better monitor germination timing within the vials. Achenes were placed inside vial tubes on two discs of filter paper and moistened with 0.6 mL of distilled water. Vials were sealed in the same way (septum plus parafilm). Ethylene concentrations were determined via gas chromatography (Hewlett Packard 4890, Palo Alto, CA, USA) using a prepacked column (Porapak N 80/100 mesh, length 2 m) and a flame ionization detector. The injector, the column, and the detector had temperatures of 110, 90, and 250 °C, respectively. All replicates were measured independently, and the analysis was conducted from day 3 to day 5 from seed imbibition. Ethylene production was determined by integrating the peaks of ethylene produced multiplied by the flow rate and normalized to the achene dry weight.

3. Results and Discussion

3.1. Effect of Constant and Alternating Temperatures on Germination

Dry after-ripened wild cardoon achenes did not germinate at a constant temperature of 10 °C and only 3 and 6% of achenes germinated at constant 15 and 20 °C (Figure 1). In contrast, the exposure to alternating temperature regimes elicited dormancy termination causing an abrupt increase in germination response up to 80% ($P < 0.001$). The effect of alternating temperatures on germination started from day 3 of imbibition onwards. Maximum germination increased to day 6 and no further germination was scored until the end of the experiment. These results are in line with the previous studies using dry after-ripened wild cardoon achenes, confirming that this treatment can be used for dormancy alleviation in this plant [24,46].

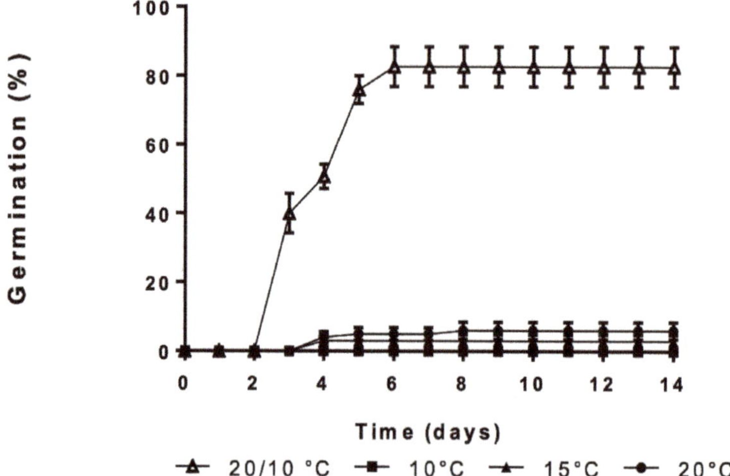

Figure 1. Germination time course percentages of wild cardoon achenes incubated at alternating temperatures (open up-pointing triangle) or constant (closed square, circle, and down-pointing triangle, respectively). Vertical bars indicate the SEs.

3.2. Seed Transcriptome Annotation

In the present study, for the first time for wild cardoon, a seed transcriptome analysis was carried out providing the transcripts composition and dynamics related to the physiological response respect to different imbibition temperatures. We obtained a total of 63,827,612 read pairs with a mean Q always above 34. The datasets generated and analyzed in the current study are available in the NCBI SRA repository PRJNA627453 (https://www.ncbi.nlm.nih.gov/bioproject/ PRJNA627453). Using all the RNA-Seq reads samples, we obtained 20,610 transcripts with an expression value > 0. With the means of the in silico functional annotation, 14,078 genes were identified belonging to 60 GO functional groups including biological process (29 subcategories), cellular component (18 subcategories) and molecular function (13 subcategories) (Figure S1). For biological process, 'cellular process' and 'metabolic process' were dominant terms, while for molecular function 'catalytic activity' and 'binding' were the major subcategories. The prominence of 'binding' term suggested a crucial role of TFs in seed germination regulation as seen for flower head development in cultivated cardoon [39] or seed transcriptome of other plant families [29]. For cellular components, the identified GO terms were more evenly spread across the subcategories with 'cell' and 'cell part' accounting for the most numerous ones. Enrichment analysis using the *C. cardunculus* transcriptome [39] as the reference confirmed the relatively higher amount of 'catalytic activity' term for molecular function, 'cell part' and 'cell' for cellular component and 'metabolic process' and 'cellular process' respect to the *C. cardunculus* transcriptome including several phenological stages (Figure S2). As for biological processes the over-representation of 'metabolic process' and 'cellular process' reveal the upregulation of pathways, including anabolism and catabolism, and communication occurring among cells. Interestingly, we found the 'signaling' and 'response to stimulus' terms enriched respect to the reference annotation. Another highly represented functional group was 'binding', which includes transcription factors activity important for seed germination. We identified transcripts, 9226 (39.0% of the total number of transcripts) accounting for 57 TF families (Figure S3). In general, the alternating temperature treatment condition exhibited the highest number of genes attributable to TF families, while the dry achenes the lowest. Considering the seed transcriptome including all the treatment conditions, the TF family with the highest number of representatives were bHLH (1026), then MYB/MYB-related (931), NAC (593) and ERF (532), which altogether represent the 33.4% of the identified TFs (Figure S3). Similar TFs families composition was observed in other species [29,30], but in the present study, we observed some TFs families usually not highly represented as GAI-RGA-SCARECROW (GRAS), FRS (FAR1 Related Sequences) and Golden2-like (G2-like) that accounted for the 8.8% of the total identified TFs.

3.3. Differential Expression Analysis

The analysis of the differentially expressed genes (DEGs) across all the samples resulted in 4737 sequences (Table S2) and their expression profiles were confirmed by qRT-PCR analysis, which showed a good correlation ($R^2 = 0.64$) with RNA-Seq data (Figure S4A,B), supporting the reliability of our dataset. The variance of expression levels across the samples confirmed the marked difference of transcriptional regulation produced by the exposition to different environmental factors such as imbibition to constant or alternating temperature (Figure S5). Enrichment analysis on DEGs showed an upregulation of functional GO terms associated with binding and catalytic processes (Figure S6). This expression pattern testifies of an intense enzymatic activity that is supported by the TFs. Similar diversity in transcriptional profile was already reported for *Paris polyphylla* seeds exposed to warm stratification respect to non-stratified seeds [30], but, to the best of our knowledge, this is the first report investigating the transcriptional profile variation during dormancy termination. To identify the most relevant functional groups involved in its regulation, we generated an expressed transcripts matrix that comprised 764 correlated DEGs, which consisted in 131 GO terms, mostly enriched for 'metabolic process', 'cellular process', 'biological regulation' and 'response to stimulus' amongst the 'biological process' GO category, while 'binding', 'activity of structural molecules', 'catalytic activity' and 'antioxidant activity' for 'molecular function' GO category (Figure S7 and Table S3). To analyze the association amongst the correlated transcripts, we generated a correlation-based network, which showed the 764 transcripts, as nodes, connected by 1344 edges, at Pearson correlation coefficient of 0.90 (Figure S8). The largest connected component of the network with 347 nodes and 943 edges is shown in Figure 2, which functional annotation is reported in Table 1 for selected genes which GO functional annotation was associated with seed dormancy, ethylene, and ROS homeostasis.

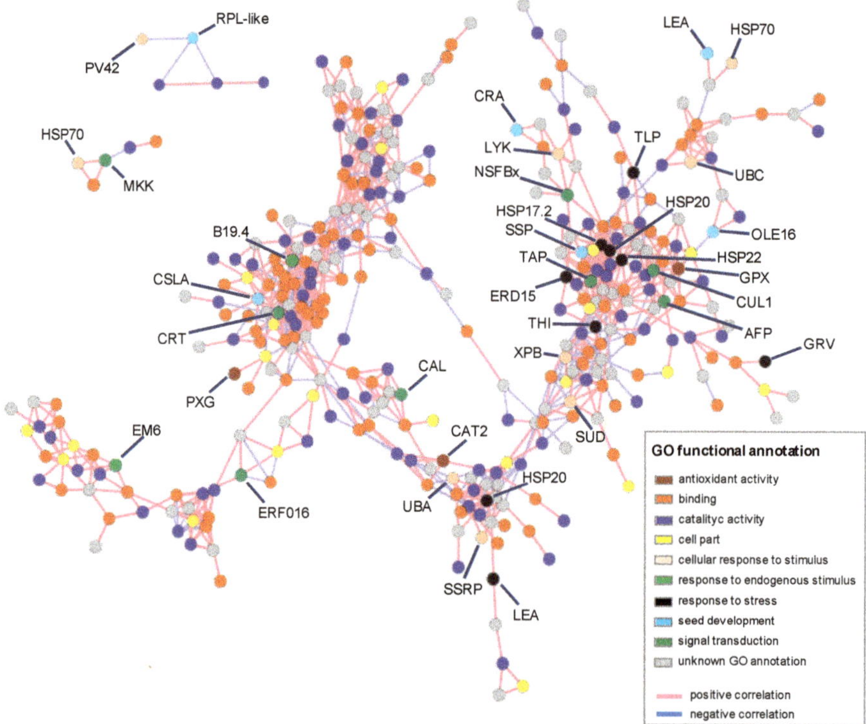

Figure 2. The layout of the three largest connected components of the correlation-based co-expression network obtained using differentially expressed genes (DEGs), in which are reported the putative gene names for antioxidant activity, cellular response to stimulus, response to endogenous stimulus, response to stress, seed development, and signal transduction GO terms. Other independent smaller-size subnetworks are shown in Figure S8.

Table 1. Gene annotation of transcripts associated with seed dormancy, ethylene, and reactive oxygen species (ROS) homeostasis.

Gene Identifier	Gene Description	Effect on Dormancy Regulation
Ccrd_v2_22156_g15	*GPX—phospholipid hydroperoxide glutathione peroxidase*	Reduction of H_2O_2 or organic hydroperoxides [47]
Ccrd_v2_22208_g15	*RBOH—Respiratory burst oxidase protein*	Biosynthesis of superoxide/Dormancy alleviation factor [20]
split_gene_Ccrd_v2_02613_g01-g146	*CAT2—Catalase-like isoform X2*	Protect from H_2O_2 and lipid peroxidation [48]
Ccrd_v2_14857_g10	*PXG—Plant seed peroxygenase*	Protect from dehydration stress [49]

Table 1. *Cont.*

Gene Identifier	Gene Description	Effect on Dormancy Regulation
Ccrd_v2_04883_g02	*DOGL3—protein DOG1-like 3 isoform*	Effect unclear
Ccrd_v2_02779_g02	*CUL1—Cullin-1-like isoform X1*	Control of ABA biosynthesis [50]
Ccrd_v2_12828_g08	*ERF9—ETHYLENE RESPONSE FACTOR9*	Ethylene signaling [51]
Ccrd_v2_23452_g16	*SPA—protein SUPPRESSOR OF PHYA-105 1-like isoform*	Regulates circadian rhythms/germination enhancer [52]
Ccrd_v2_21316_g15	*MKK—Mitogen-activated protein kinase 9-like*	Induces the synthesis of ethylene [53]
Ccrd_v2_08680_g05	*NSFBx—Probable F-box protein (At5g04010)*	Effect unknown
Ccrd_v2_00002_g01	*TAP—2A phosphatase associated protein*	Dormancy enhancer [54]
Ccrd_v2_09661_g05	*AFP—ninja-family protein AFP3-like*	Control of ABA biosynthesis [55]
Ccrd_v2_03046_g02	*ERF016—Ethylene-responsive transcription factor ERF016-like*	Effect unclear
novel_gene_1_5b8548b9	*LTI65—low-temperature-induced 65 kDa protein-like*	Responsive to ABA [56]
Ccrd_v2_06887_g03	*PV42—SNF1-related protein kinase regulatory subunit gamma-like PV42a*	Dormancy enhancer [57]
Ccrd_v2_15915_g11	*PARP—putative Poly [ADP-ribose] polymerase 3*	DNA protection system
Ccrd_v2_01121_g01	*UBC—Ubiquitin-conjugating enzyme E2 2*	Induced by ABA [58]
Ccrd_v2_24782_g17	*ACO1—1-AMINOCYCLOPROPANE-1-CARBOXYLATE OXIDASE1*	Ethylene biosynthesis [51]
Ccrd_v2_23833_g17	*ACO4—1-AMINOCYCLOPROPANE-1-CARBOXYLATE OXIDASE4*	Ethylene biosynthesis [51]
Ccrd_v2_16461_g11	*ACS10—ACC synthase10*	Ethylene biosynthesis [51]
Ccrd_v2_19164_g13	*LYK—LYSIN MOTIF RECEPTOR KINASE*	Effect unknown
Ccrd_v2_01115_g01	*SSRP—FACT complex subunit SSRP1-like isoform X1*	Dormancy enhancer [59]
Ccrd_v2_14002_g09	*UBA—ubiquitin-activating enzyme E1 1-like isoform X*	Effect unknown
Ccrd_v2_22183_g15	*XPB—general transcription and DNA repair factor IIH helicase subunit XPB1*	DNA repair [60]
Ccrd_v2_00258_g01	*ERD15—protein EARLY RESPONSIVE TO DEHYDRATION 15-like*	Induced by dehydration stress/Modulates ABA response [61]
Ccrd_v2_22449_g15	*CYP707A2—Cytochrome P450, Family 707, Subfamily A, Polypeptide2*	Reduced dormancy [51]
Ccrd_v2_15609_g11	*NCED9—NINE-cis-EPOXYCAROTENOID DIOXYGENASE -9*	Dormancy enhancer [51]
Ccrd_v2_13522_g09	*ABI5—protein ABSCISIC ACID-INSENSITIVE 5*	Dormancy enhancer [55]
Ccrd_v2_00305_g01	*EM6—em-like protein GEA6*	Effect unclear
Ccrd_v2_02516_g01	*TLP—thaumatin-like protein 1b*	Effect unknown
Ccrd_v2_05395_g03	*THI—Gamma thionin*	Effect unknown
Ccrd_v2_00955_g01	*CSLA—CELLULOSE SYNTHASE-LIKE (CSL)*	Effect unknown
Ccrd_v2_08837_g05	*RPL-like—Ribosomal Protein-like*	Reduced dormancy [62]
Ccrd_v2_19371_g13	*EXPA6—Expansin A6*	Cell wall loosening/Ethylene signaling [51]
Ccrd_v2_05454_g03	*EXPA11—Expansin A11*	Cell wall loosening/Ethylene signaling [51]
Ccrd_v2_19070_g13	*EXPA11—Expansin A11*	Cell wall loosening/Ethylene signaling [51]
Ccrd_v2_21520_g15	*EXPA1—Expansin A1*	Cell wall loosening/Ethylene signaling [51]
Ccrd_v2_01833_g01	*EXPA1-like—Expansin A11-like*	Cell wall loosening/Ethylene signaling [51]
Ccrd_v2_00414_g01	*EXPA6—Expansin A6*	Cell wall loosening/Ethylene signaling [51]

Binding and catalytic activity GO terms were uniformly spread across the network. On the contrary, some putative transcripts encoding for response to ROS stress as *GLUTATHIONE PEROXIDASE (GPX)*, reported being upregulated in the presence of oxidative stress [47], or as *HEAT-SHOCK PROTEINS (HSPs)* [63] were not uniformly distributed across the network, and formed a closely associated cluster. Within this group, we observed the presence of a cullin protein (*CUL1*), the family of which is associated with degradation of *ABA INSENSITIVE 5 (ABI5)* [50] and *ABI FIVE BINDING PROTEIN (AFP)* that participates in the control of *ABI5* accumulation [55]. This tight association can suggest a close interaction of these transcripts in the removal of the last dormancy constraints. On the other hand, this group also included *TYPE 2A PHOSPHATASE-ASSOCIATED PROTEIN 46*

(TAP46), which is known to stabilize *ABI5* transcript expression [54]. Therefore, the nature of their interaction should be further investigated to unveil how they can modulate ABA levels. Another smaller cluster consisted of cardoon homologous to *PV42* (an *SNF1*-related protein kinase regulatory subunit gamma-like), a *RIBOSOMAL PROTEIN L-LIKE (RPL-LIKE)*, and three catalytic genes. *SNF1* and *RPL* interaction was associated with the completion of germination in Arabidopsis seeds [64]. Moreover, *MITOGEN-ACTIVATED PROTEIN KINASE 9-like (MKK)*, which is associated with the induction of ethylene synthesis [53], was not included in the main network and the annotation of closely linked genes is not sufficiently clear for this plant species to speculate for a possible interaction among them. Similarly, other cardoon putative transcripts associated with ethylene such as *RESPONSIVE TRANSCRIPTION FACTORS (ERF)*, *1-AMINOCYCLOPROPANE-1-CARBOXYLIC ACID OXIDASE (ACO)*, *AMINOCYCLOPROPANE-1-CARBOXYLIC ACID SYNTHASE (ACS)* or with the modulation of ROS, such as *RESPIRATORY BURST OXIDASE PROTEIN (RBOH)* were not included in the main subnetwork.

However, when we analyzed the transcriptional dynamics of a broader set of genes (Figure 3), we observed an increase of dehydration stress response as *HSPs*, *PEROXIDASE (GPX)*, *PEROXYGENASE (PXG)*, and *LATE EMBRYOGENESIS ABUNDANT (LEA)* in dry achenes. Moreover, in this condition, there were a richer composition of *SEED STORAGE PROTEINS (SSP)*, Oleosin proteins and DNA repair system factors, such as *XPB*. These findings are in line with transcripts composition recently described for dry *A. thaliana* seeds [64]. However, at a constant temperature, the transcriptional program changed abruptly with the expression of homologous genes associated with ABA signaling and biosynthesis, as can be drawn from the upregulation of *ABI5*, *TAP*, *PV42* and *LTI65* transcripts expression. These results suggest increased ABA biosynthesis at a constant temperature, which is supported by the upregulation of *NCED9* at this condition. Respect to dry seed, at the constant temperature we observed a downregulation of *DELAY OF DORMANCY-LIKE3 (DOGL3)*. This expression is further dramatically reduced at alternating temperatures. Recently, Sall et al. [64] reported that the overexpression of *DOGL3* caused ABA hypersensitivity in seed germination of *A. thaliana* but proposed a role as an inducer of seed reserve accumulation for *DOGL* genes respect to dormancy modulator which characterizes *DOG1*. Further research is needed to confirm whether *de novo* ABA biosynthesis associated with exposure to constant temperature is the main mechanism involved in the dormancy maintenance of wild cardoon. The alternating temperature condition, instead, stimulated the expression of *RBOH* that is responsible for the biosynthesis of superoxide and was associated with dormancy alleviation in sunflower [20]. The differential expression of some *CATALASE* family genes can support the presence of an oxidation stress control system acting differently depending on the physiological step. *CYP707A2* associated with ABA degradation [51] and *RPL* reported as a stimulator of germination completion [64]. Moreover, transcripts encoding for ethylene metabolism were upregulated, such as *ACO1*, *ACO4*, and emphACS10, and also signaling with *EXPANSINs*. The latter is responsible for plant cell wall loosening through ethylene promotion of micropylar endosperm weakening by inducing the expression of *CELL WALL REMODELLING PROTEINS (CWRP)* and/or ROS that cause cell wall loosening or cell separation of this tissue [65]. The upregulation of *CELLULOSE SYNTHASE-LIKE A (CSLA)* at the same imbibition condition can probably be reconducted to this reorganization of plant cell walls for germination completion. Whether the stimulation of ROS signaling and ethylene biosynthesis and signaling has a major role in dormancy termination of wild cardoon needs to be confirmed by further studies unveiling the interaction among their pathways.

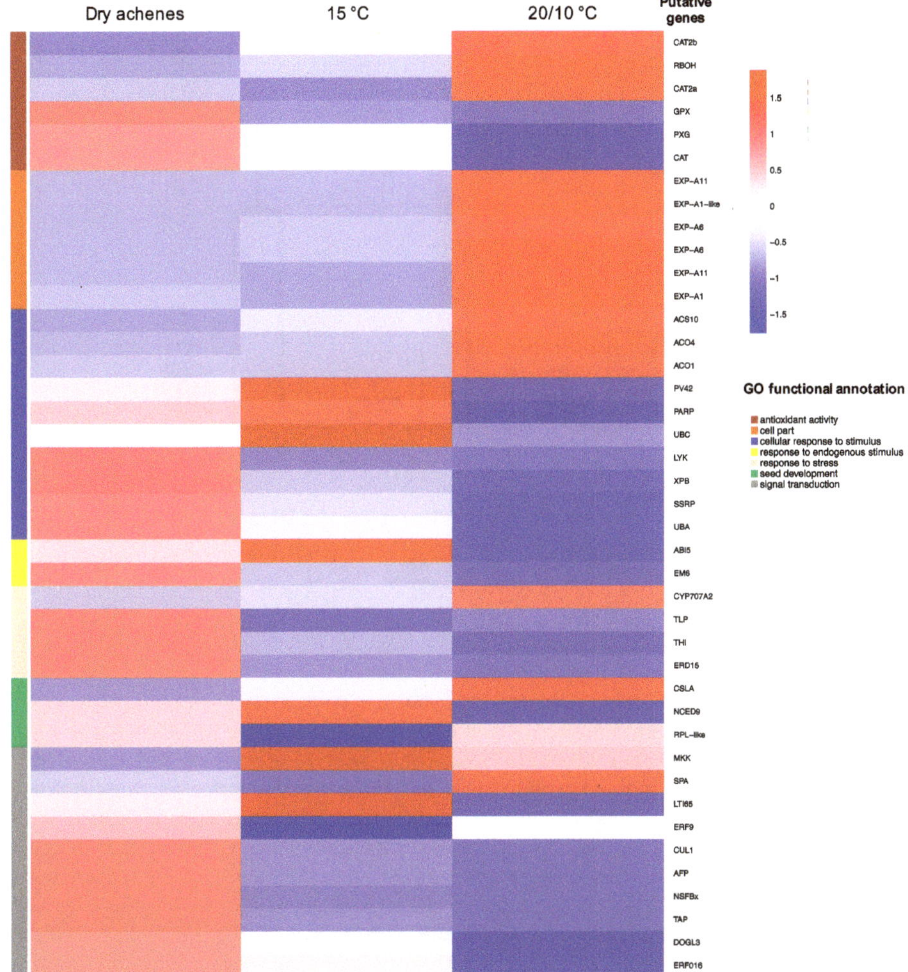

Figure 3. Expression heatmap of a set of transcripts associated with seed dormancy, ethylene, and ROS homeostasis in dry achenes, imbibed achenes at constant temperature (15 °C) and imbibed achenes at alternating temperatures (20/10 °C). The color scale represents the log2-transformed TPM value.

3.4. Effect of Incubation Temperature on Ethylene Synthesis and Germination

Ethylene production was detected one day before the beginning of germination and its content was always significantly higher at alternating temperatures in comparison with that scored at constant temperature (Figure 4). Ethylene content at day 3 was 0.31 and 0.06 for 20/10 °C and 15 °C, respectively. It gradually increased at alternating temperatures to reach its maximal value at day 5 of incubation (0.78 ppm). In contrast, at the same time, just 0.23 ppm was measured at constant temperatures. Similar results on ethylene production during seed germination were previously published [17].

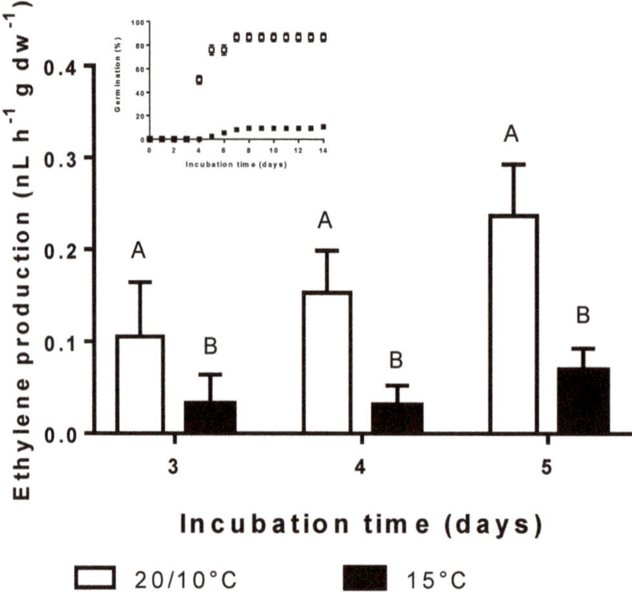

Figure 4. Ethylene production (±SE) in achenes exposed to constant 15 °C (closed bars) or alternating temperatures 20/10 °C (open bars) for 5 days.

Moreover, the important role of ethylene in germination completion was also supported by the results obtained from the use of inhibitors of its biosynthesis and signaling. In all cases, germination at alternating temperatures was gradually reduced (Figure 5A–D). For instance, the presence of AIB, an inhibitor of ACC oxidase, reduced germination at alternating temperatures to 8 % being statistically comparable to germination scored at constant temperature of the control (Figure 5A) ($P < 0.05$). Similarly, $CoCL_2$ at a dose of 1.25 and 2.5 mM lowered significantly total germination at alternating temperature, compared to the constant temperature of the control (Figure 5B). Also, the interference on ethylene receptors had a significant negative effect on dormancy termination by alternating temperatures (Figure 5C,D). The use of 0.25–1 mM $AgNO_3$ and 25–100 µm 1-MCP inhibit the germination up to 50%, respectively. On the other hand, the germination was only partially increased in the presence of ethephon, an ethylene releasing compound, (Figure 5E,F). At alternating temperature, ethephon increased germination to 91%, but at constant temperatures, the germination response was lower than 40%. That is, the ethylene alone did not replace the requirement of alternating temperatures for dormancy termination. Germinations scored at constant temperatures by the use of ethephon agree with that reported by Kepczynski and Sznigir [66] using 16 weeks dry after-ripened *Amaranthus retroflexus* seeds and similar to Corbineau et al. [18]. Ethephon replaces the requirement of dormancy alleviation factors, such as cold stratification and dry after ripening, in several species presenting primary dormancy [18]. Thus, these results led us to suggest that alternating temperatures elicits ethylene biosynthesis and probably signaling as well. However, the increase of ethylene production does not necessarily imply that its presence can determine dormancy termination in wild cardoon. This hormone plays an important role in the modulation of last dormancy constraint in concert with other key players, such as ABA/GA and/or ROS [18]. Further research is needed to ascertain whether this is the major mechanism for ABA/GA balance modulation able to produce a reduced ABA sensitivity, as previously described in the presence of alternating temperatures [12,37].

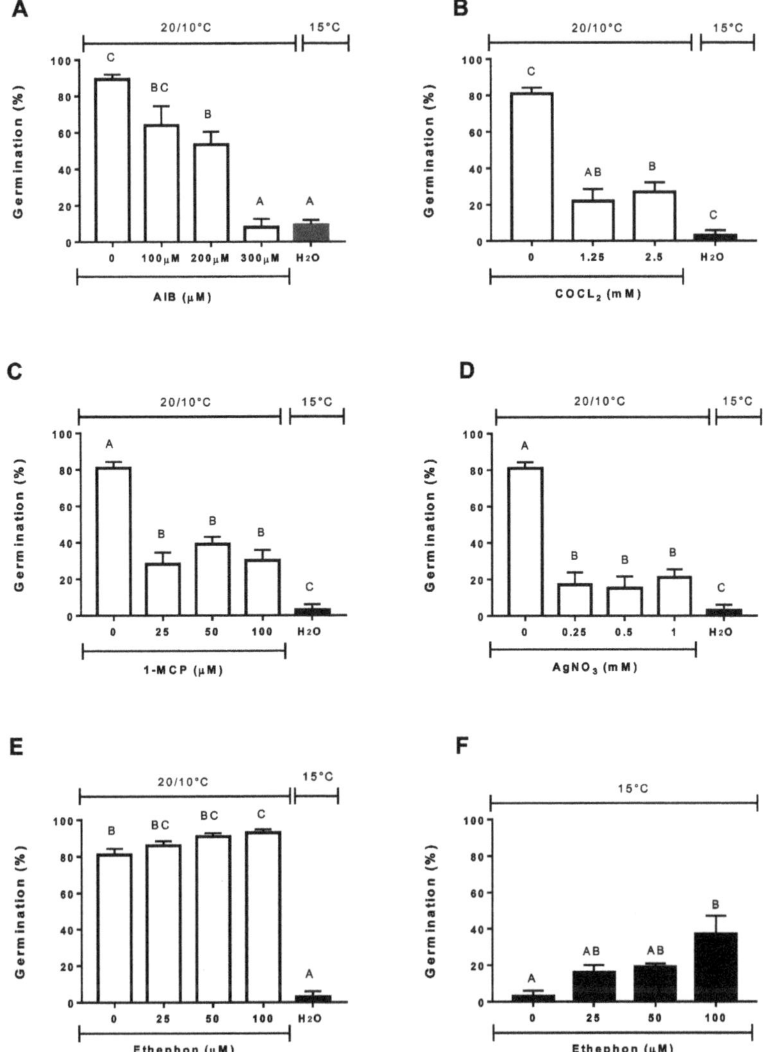

Figure 5. Final germination percentages of wild cardoon achenes incubated at alternating temperatures (20/10 °C) (open bar) or constant (15 °C) in the presence of ethylene biosynthesis inhibitors (**A,B**), inhibitors of ethylene binding to its receptor (**C,D**), or ethylene donor (**E,F**). Vertical bars indicate the SEs. Similar letters at the top of each bar indicate no differences according to Tukey's test ($P < 0.05$).

3.5. Effect of Antioxidants, ROS Scavengers, and ROS Donors on Germination at Alternating or Constant Temperatures

The presence of antioxidants, such as ascorbic acid and glutathione, progressively reduced significant germination in achenes exposed to alternating temperatures determining similar percentages observed at constant temperature of the control (Figure 6A,B). This suggests that dormancy termination by alternating temperatures could require the presence of ROS compounds. To support this hypothesis, further germination experiments were carried out in the presence of DMTU, a hydrogen peroxide scavenger, and DPI, a ROS synthesis inhibitor. Both compounds reduced germination to an extent similar to those obtained at constant temperature of the control ($P < 0.001$) (Figure 6C,D). Results obtained for DPI and DMTU are in agreement with those published by Leymarie et al. [20] using *A. thaliana* seeds.

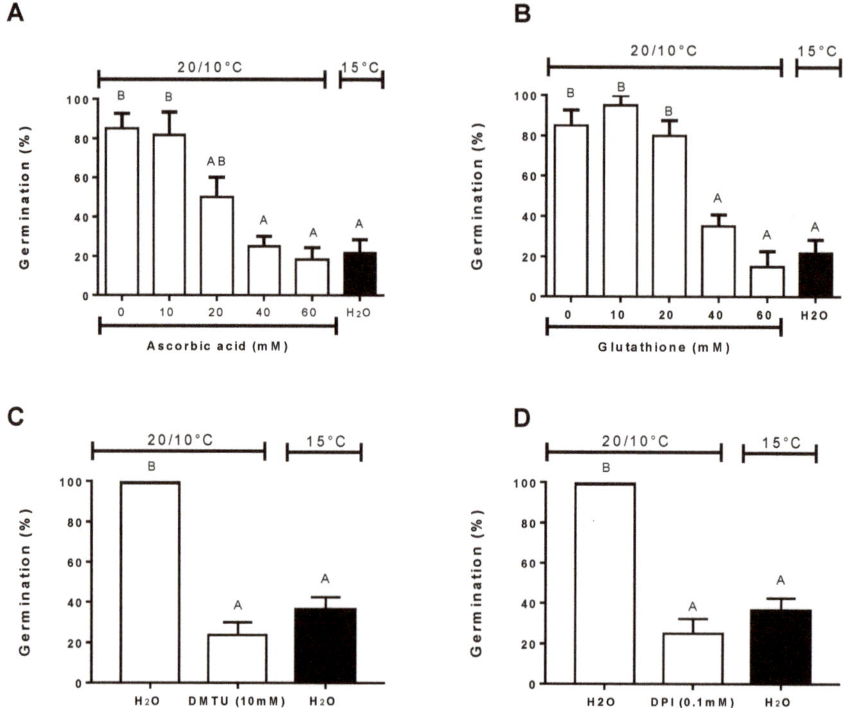

Figure 6. Final germination percentages of wild cardoon achenes incubated at alternating temperatures (20/10 °C) (open bar) or constant (15 °C) in the presence of antioxidant compounds (**A,B**), hydrogen peroxide scavenger (**C**), and NAD(P)H oxidase inhibitors (**D**). Vertical bars indicate the SEs. Similar letters at the top of each bar indicate no differences according to Tukey's test ($P < 0.05$).

On the other hand, at constant temperature, the use of methyl viologen, a ROS donor, significantly increased germination from 8 to 57% at doses of 0 and 0.5 mM, respectively (Figure 7A), similarly to previous investigations in sunflower achenes [23,67]. Likewise, the inhibition of peroxidase, an enzyme that reduces the amount of hydrogen peroxide, in the presence of SHAM, partially increased germination at constant a temperature (Figure 7B). Although the results obtained in the present study have to be confirmed by direct ROS measurements, altogether these findings suggest that dormancy termination by alternating temperatures may include the involvement of ROS compounds

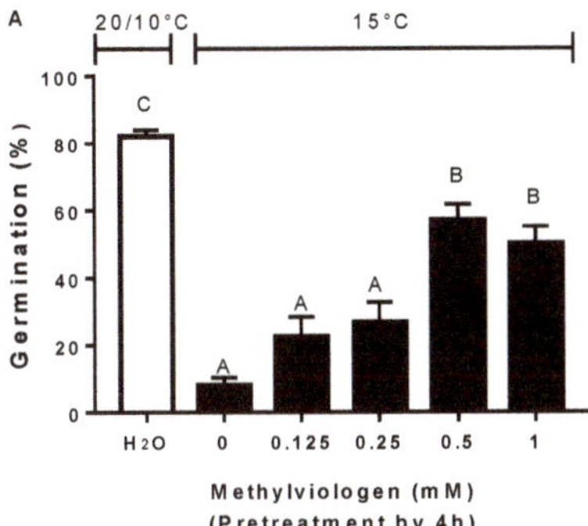
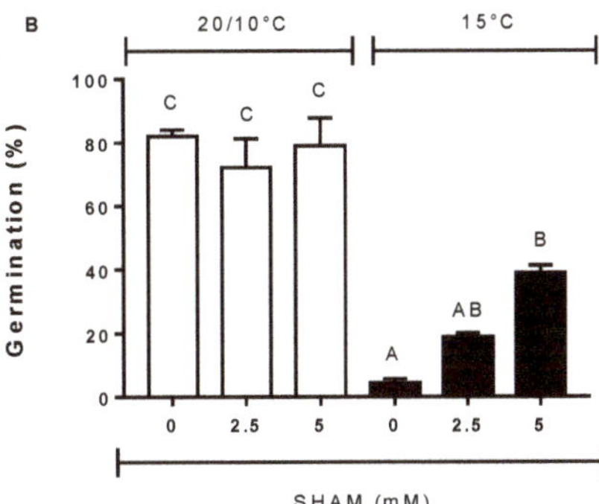

Figure 7. Final germination percentages of wild cardoon achenes incubated at alternating temperatures (20/10 °C) (open bar) or constant (15 °C) in the presence of methyl viologen (**A**) and salicylhydroxamic acid (SHAM) (**B**). Vertical bars indicate the SEs. Similar letters at the top of each bar indicate no differences according to Tukey's test ($P < 0.05$).

4. Conclusions

The present study is the first large-scale gene expression investigation on dormancy termination process in wild cardoon. Transcriptome patterns associated with the imbibition at constant temperature include upregulation of ABA biosynthesis genes, ABA-responsive genes, as well as other genes previously related to physiological dormancy and inhibition of germination. While expression patterns stimulated at alternating temperatures comprised ethylene and ROS signaling and metabolism together with ABA degradation and cell wall loosening. Physiological assays support molecular data showing that ethylene is necessary for dormancy termination at alternating temperatures, even if its presence does not imply the completion of germination. Similarly, ROS is needed for dormancy termination since its depletion hampers this process, but ROS donors cannot overcome dormancy completely. These findings suggest an important role of both ethylene and ROS in dormancy termination at alternating temperatures, most probably as a fine-tuned mechanism for environmental sensing. This can be a very useful system for effectively achieving dormancy termination once environmental conditions are suitable for germination in highly disturbed habitat in which this plant lives. Moreover, our results may have applications in naturalization efforts using wild cardoon for the naturalization of highly disturbed habitats impacted by human activity (e.g., sowing seeds at the correct environmental temperature regimes).

Supplementary Materials: The following are available online at http://www.mdpi.com/2223-7747/9/9/1225/s1, Table S1. qRT-PCR primers designed on selected transcripts identified within the transcriptome assembly and used for RNA-seq data validation. Table S2. Expression levels, reported as log2-transformed TPM value, of 4737 DEGs identified in this study. Table S3. Complete list of correlated transcripts present in the co-expression network analysis. Figure S1. In silico functional annotation of identified transcripts. Gene Ontology (GO) terms are reported for each GO category. Figure S2. Enrichment analysis of seed transcriptome annotation compared to the *C. cardunculus* transcriptome including several phenological stages. Figure S3. Amount of transcription factor families identified in this study. Figure S4. qRT-PCR analysis for RNA-Seq data validation. A: Relative expression of wild cardoon homologous transcripts associated with regulation of ethylene, ROS and ABA. The relative expression ratio is expressed as the fold increase relative to dry achenes. The error bars represent the standard error of the mean of three biological replicates. Letters indicate significantly different values according to ANOVA (p-value ≤ 0.05). B: Correlation of gene expression results obtained from real-time PCR analysis and RNA-Seq (TPM) for 6 selected genes across samples. The correlation of determination (R2) was 0.64. Figure S5. Hierarchical cluster analysis of differentially expressed genes in dry achenes, imbibed achenes at a constant temperature (15 °C) and imbibed achenes at alternating temperatures (20/10 °C). The color scale represents the log2-transformed TPM value. Figure S6. Enrichment analysis of the 764 highly correlated DEGs. Figure S7. Expression heatmap of the 764 highly correlated DEGs. DS: dry achenes; IDS: imbibed achenes at constant temperature (15 °C); INDS: imbibed achenes at alternating temperatures (20/10 °C). The color scale represents the log2-transformed TPM value. Figure S8. The layout of all the correlation-based co-expression networks obtained using of the 764 highly correlated DEGs, in which are functional GO categories.

Author Contributions: G.D.P., H.R.H. and S.A.R. conceived and designed the study; G.D.P. and H.R.H. collected the samples; G.D.P. and H.R.H. conducted the laboratory experiments; A.D.P. and G.D.P. conducted the bioinformatic analyses; G.D.P., H.R.H. and A.D.P. wrote the manuscript. S.A.R. contributed to the experimental design and writing of the manuscript. All authors have read and agreed to the published version of the manuscript.

Funding: This work was financed with funding provided by Catania section of the Istituto per i Sistemi Agricoli e Forestali del Mediterraneo (CNR-ISAFOM) and by (i) LomasCyT IV (FCA-025), Universidad Nacional de Lomas de Zamora, Llavallol, Argentina; (ii) PICT 2015-3087 from ANPCyT, Argentina. Bioinformatic analyses were supported by St. Petersburg University (project ID #51555422).

Acknowledgments: We thank Pietro Calderaro for his technical assistance in the field operations and Elisa Miraglia for her contribution in bioinformatic analyses. We acknowledge the support of ELIXIR-IT, CINECA for the provision of computational resources service.

Conflicts of Interest: The authors declare no conflict of interest.

References

1. Finch-Savage, W.E.; Leubner-Metzger, G. Seed dormancy and the control of germination. *New Phytol.* **2006**, *171*, 501–523. [CrossRef] [PubMed]
2. Long, R.L.; Gorecki, M.J.; Renton, M.; Scott, J.K.; Colville, L.; Goggin, D.E.; Commander, L.E.; Westcott, D.A.; Cherry, H.; Finch-Savage, W.E. The ecophysiology of seed persistence: A mechanistic view of the journey to germination or demise. *Biol. Rev.* **2015**, *90*, 31–59. [CrossRef] [PubMed]
3. Finch-Savage, W.E.; Footitt, S. Seed dormancy cycling and the regulation of dormancy mechanisms to time germination in variable field environments. *J. Exp. Bot.* **2017**, *68*, 843–856. [CrossRef] [PubMed]
4. Batlla, D.; Benech-Arnold, R.L. A framework for the interpretation of temperature effects on dormancy and germination in seed populations showing dormancy. *Seed Sci. Res.* **2015**, *25*, 1–12. [CrossRef]
5. Chao, W.S.; Foley, M.E.; Dogramaci, M.; Anderson, J.V.; Horvath, D.P. Alternating temperature breaks dormancy in leafy spurge seeds and impacts signaling networks associated with HY5. *Funct. Integr. Genom.* **2011**, *11*, 637–649. [CrossRef]
6. Puglia, G.; Carta, A.; Bizzoca, R.; Toorop, P.; Spampinato, G.; Raccuia, S.A. Seed dormancy and control of germination in *Sisymbrella dentata* (l.) O.E. Schulz (Brassicaceae). *Plant Biol.* **2018**, *50*, 164–170. [CrossRef]
7. Bewley, J.D.; Bradford, K.J.; Hilhorst, H.W.M.; Nonogaki, H. Dormancy and the Control of Germination. In *Seeds: Physiology of Development, Germination and Dormancy*, 3rd ed.; Springer: New York, NY, USA, 2013; pp. 247–297. ISBN 978-1-4614-4693-4.
8. Vandelook, F.; Newton, R.J.; Carta, A. Photophobia in lilioid monocots: Photoinhibition of seed germination explained by seed traits, habitat adaptation and phylogenetic inertia. *Ann. Bot.* **2018**, *121*, 405–413. [CrossRef]
9. Saatkamp, A.; Affre, L.; Dutoit, T.; Poschlod, P. Germination traits explain soil seed persistence across species: The case of Mediterranean annual plants in cereal fields. *Ann. Bot.* **2011**, *107*, 415–426. [CrossRef]
10. Picciau, R.; Pritchard, H.W.; Mattana, E.; Bacchetta, G. Thermal thresholds for seed germination in Mediterranean species are higher in mountain compared with lowland areas. *Seed Sci. Res.* **2019**, *29*, 44–54. [CrossRef]
11. Arana, M.V.; Tognacca, R.S.; Estravis-Barcalá, M.; Sánchez, R.A.; Botto, J.F. Physiological and molecular mechanisms underlying the integration of light and temperature cues in *Arabidopsis thaliana* seeds. *Plant Cell Environ.* **2017**, *40*, 3113–3121. [CrossRef]
12. Huarte, H.R.; Luna, V.; Pagano, E.A.; Zavala, J.A.; Benech-Arnold, R.L. Fluctuating temperatures terminate dormancy in *Cynara cardunculus* seeds by turning off ABA synthesis and reducing ABA signaling, but not stimulating GA synthesis or signaling. *Seed Sci. Res.* **2014**, *24*, 79–89. [CrossRef]
13. Laspina, N.V.; Batlla, D.; Benech-Arnold, R.L. Dormancy cycling in *Polygonum aviculare* seeds is accompanied by changes in ABA sensitivity. *J. Exp. Bot.* **2020**. [CrossRef] [PubMed]
14. Ghassemian, M.; Nambara, E.; Cutler, S.; Kawaide, H.; Kamiya, Y.; McCourt, P.; Corbineau, F.; Strnad, M.; Lynn, J.R.; Finch-Savage, W.E.; et al. Regulation of abscisic acid signaling by the ethylene response pathway in Arabidopsis. *Plant Cell* **2000**, *12*, 1117–1126. [CrossRef]
15. Linkies, A.; Müller, K.; Morris, K.; Turecková, V.; Wenk, M.; Cadman, C.S.C.; Corbineau, F.; Strnad, M.; Lynn, J.R.; Finch-Savage, W.E.; et al. Ethylene interacts with abscisic acid to regulate endosperm rupture during germination: A comparative approach using *Lepidium sativum* and *Arabidopsis thaliana*. *Plant Cell* **2009**, *21*, 3803–3822. [CrossRef] [PubMed]

16. Kucera, B.; Cohn, M.A.; Leubner-Metzger, G. Plant hormone interactions during seed dormancy release and germination. *Seed Sci. Res.* **2005**, *15*, 281–307. [CrossRef]

17. Matilla, A.J.; Matilla-Vázquez, M.A. Involvement of ethylene in seed physiology. *Plant Sci.* **2008**, *175*, 87–97. [CrossRef]

18. Corbineau, F.; Xia, Q.; Bailly, C.; El-Maarouf-Bouteau, H. Ethylene, a key factor in the regulation of seed dormancy. *Front. Plant Sci.* **2014**, *5*, 539. [CrossRef] [PubMed]

19. El-Maarouf-Bouteau, H.; Bailly, C. Oxidative signaling in seed germination and dormancy. *Plant Signal. Behav.* **2008**, *3*, 175–182. [CrossRef] [PubMed]

20. Leymarie, J.; Vitkauskaité, G.; Hoang, H.H.; Gendreau, E.; Chazoule, V.; Meimoun, P.; Corbineau, F.; El-Maarouf-Bouteau, H.; Bailly, C. Role of reactive oxygen species in the regulation of Arabidopsis seed dormancy. *Plant Cell Physiol.* **2012**, *53*, 96–106. [CrossRef] [PubMed]

21. Bailly, C. The signaling role of ROS in the regulation of seed germination and dormancy. *Biochem. J.* **2019**, *476*, 3019–3032. [CrossRef]

22. Oracz, K.; El-Maarouf-Bouteau, H.; Kranner, I.; Bogatek, R.; Corbineau, F.; Bailly, C. The mechanisms involved in seed dormancy alleviation by hydrogen cyanide unravel the role of reactive oxygen species as key factors of cellular signaling during germination. *Plant Physiol.* **2009**, *150*, 494–505. [CrossRef] [PubMed]

23. El-Maarouf-Bouteau, H.; Sajjad, Y.; Bazin, J.; Langlade, N.; Cristescu, S.M.; Balzergue, S.; Baudouin, E.; Bailly, C. Reactive oxygen species, abscisic acid and ethylene interact to regulate sunflower seed germination. *Plant Cell Environ.* **2015**, *38*, 364–374. [CrossRef] [PubMed]

24. Huarte, H.R.; Borlandelli, F.; Varisco, D.; Batlla, D. Understanding dormancy breakage and germination ecology of *Cynara cardunculus* (Asteraceae). *Weed Res.* **2018**, *58*, 450–462. [CrossRef]

25. Huarte, H.R.; Puglia, G.D.; Varisco, D.; Pappalardo, H.; Calderaro, P.; Toscano, V.; Raccuia, S.A. Effect of reactive oxygen species on germination of *Cynara cardunculus* (L.) cultivars. *Acta Hort.* **2020**, *2*, 33–40. [CrossRef]

26. Bailly, C.; El-Maarouf-Bouteau, H.; Corbineau, F. From intracellular signaling networks to cell death: The dual role of reactive oxygen species in seed physiology. *Comptes Rendus Biol.* **2008**, *331*, 806–814. [CrossRef]

27. Bazin, J.; Langlade, N.; Vincourt, P.; Arribat, S.; Balzergue, S.; El-Maarouf-Bouteau, H.; Bailly, C. Targeted mRNA oxidation regulates sunflower seed dormancy alleviation during dry after-ripening. *Plant Cell* **2011**, *23*, 2196–2208. [CrossRef]

28. Layat, E.; Leymarie, J.; El-Maarouf-Bouteau, H.; Caius, J.; Langlade, N.; Bailly, C. Translatome profiling in dormant and nondormant sunflower (*Helianthus annuus*) seeds highlights post-transcriptional regulation of germination. *New Phytol.* **2014**, *204*, 864–872. [CrossRef]

29. Han, Z.; Wang, B.; Tian, L.; Wang, S.; Zhang, J.; Guo, S.L.; Zhang, H.; Xu, L.; Chen, Y. Comprehensive dynamic transcriptome analysis at two seed germination stages in maize (*Zea mays* L.). *Physiol. Plant* **2020**, *168*, 205–217. [CrossRef]

30. Liao, D.; Zhu, J.; Zhang, M.; Li, X.; Sun, P.; Wei, J.; Qi, J. Comparative transcriptomic analysis reveals genes regulating the germination of morphophysiologically dormant *Paris polyphylla* seeds during a warm stratification. *PLoS ONE* **2019**, *14*, e0212514. [CrossRef]

31. Visscher, A.M.; Yeo, M.; Castillo-Lorenzo, E.; Toorop, P.E.; Junio da Silva, L.; Yeo, M.; Pritchard, H.W. *Pseudophoenix ekmanii* (Arecaceae) seeds at suboptimal temperature show reduced imbibition rates and enhanced expression of genes related to germination inhibition. *Plant Biol.* **2020**. [CrossRef]

32. Foley, M.E.; Anderson, J.V.; Chao, W.S.; Doğramacı, M.; Horvath, D.P. Initial changes in the transcriptome of *Euphorbia esula* seeds induced to germinate with a combination of constant and diurnal alternating temperatures. *Plant Mol. Biol.* **2010**, *73*, 131–142. [CrossRef]

33. Ma, W.; Guan, X.; Li, J.; Pan, R.; Wang, L.; Liu, F.; Ma, H.; Zhu, S.; Hu, J.; Ruan, Y.L.; et al. Mitochondrial small heat shock protein mediates seed germination via thermal sensing. *Proc. Natl. Acad. Sci. USA* **2019**, *116*, 4716–4721. [CrossRef] [PubMed]

34. Raccuia, S.A.; Melilli, M.G. Biomass and grain oil yields in Cynara cardunculus L. genotypes grown in a Mediterranean environment. *Field Crop Res.* **2007**, *101*, 187–197. [CrossRef]

35. Pappalardo, H.D.; Toscano, V.; Puglia, G.D.; Genovese, C.; Raccuia, S.A. *Cynara cardunculus* L. as a multipurpose crop for plant secondary metabolites production in marginal stressed lands. *Front. Plant Sci.* **2020**, *11*, 240. [CrossRef] [PubMed]

36. Raccuia, S.A.; Mainolfi, A.; Mandolino, G.; Melilli, M.G. Genetic diversity in *Cynara cardunculus* revealed by AFLP markers: Comparison between cultivars and wild types from Sicily. *Plant Breed.* **2004**, *123*, 280–284. [CrossRef]

37. Huarte, H.R.; Benech-Arnold, R.L. Hormonal nature of seed responses to fluctuating temperatures in *Cynara cardunculus* (L.). *Seed Sci. Res.* **2010**, *20*, 39–45. [CrossRef]

38. Scaglione, D.; Reyes-Chin-Wo, S.; Acquadro, A.; Froenicke, L.; Portis, E.; Beitel, C.; Tirone, M.; Mauro, R.; Lo Monaco, A.; Mauromicale, G.; et al. The genome sequence of the outbreeding globe artichoke constructed de novo incorporating a phase-aware low-pass sequencing strategy of F1 progeny. *Sci. Rep.* **2016**, *6*, 19427. [CrossRef]

39. Puglia, G.D.; Prjibelski, A.D.; Vitale, D.; Bushmanova, E.; Schmid, K.J.; Raccuia, S.A. Hybrid transcriptome sequencing approach improved assembly and gene annotation in *Cynara cardunculus* (L.). *BMC Genom.* **2020**, *21*, 317. [CrossRef] [PubMed]

40. Bolger, A.M.; Lohse, M.; Usadel, B. Trimmomatic: A flexible trimmer for Illumina sequence data. *Bioinformatics* **2014**, *30*, 2114–2120. [CrossRef]

41. Kim, D.; Langmead, B.; Salzberg, S.L. HISAT: A fast spliced aligner with low memory requirements. *Nat. Methods* **2015**, *12*, 357–360. [CrossRef]

42. Liao, Y.; Smyth, G.K.; Shi, W. FeatureCounts: An efficient general purpose program for assigning sequence reads to genomic features. *Bioinformatics* **2014**, *30*, 923–930. [CrossRef] [PubMed]

43. Götz, S.; García-Gómez, J.M.; Terol, J.; Williams, T.D.; Nagaraj, S.H.; Nueda, M.J.; Robles, M.; Talón, M.; Dopazo, J.; Conesa, A. High-throughput functional annotation and data mining with the Blast2GO suite. *Nucleic Acids Res.* **2008**, *36*, 3420–3435. [CrossRef] [PubMed]

44. Anders, S.; Pyl, P.T.; Huber, W. HTSeq-A Python framework to work with high-throughput sequencing data. *Bioinformatics* **2015**, *31*, 166–169. [CrossRef] [PubMed]

45. Rau, A.; Maugis-Rabusseau, C. Transformation and model choice for RNA-seq co-expression analysis. *Brief. Bioinform.* **2018**, *19*, 425–436. [CrossRef] [PubMed]

46. Raccuia, S.A.; Puglia, G.; Pappalardo, H.; Argento, S.; Leonardi, C.; Calderaro, P.; Melilli, M.G. Dormancy-related genes isolation in *Cynara cardunculus* var. *sylvestris*. *Acta Hortic.* **2016**, *1147*, 315–322. [CrossRef]

47. Passaia, G.; Queval, G.; Bai, J.; Margis-Pinheiro, M.; Foyer, C.H. The effects of redox controls mediated by glutathione peroxidases on root architecture in Arabidopsis. *J. Exp. Bot.* **2014**, *65*, 1403–1413. [CrossRef]

48. Bailly, C. Active oxygen species and antioxidants in seed biology. *Seed Sci. Res.* **2004**, *14*, 93–107. [CrossRef]

49. Rahman, F.; Hassan, M.; Rosli, R.; Almousally, I.; Hanano, A.; Murphy, D.J. Evolutionary and genomic analysis of the caleosin/peroxygenase (CLO/PXG) gene/ protein families in the Viridiplantae. *PLoS ONE* **2018**, *13*, e0196669. [CrossRef]

50. Jin, D.; Wu, M.; Li, B.; Bücker, B.; Keil, P.; Zhang, S.; Li, J.; Kang, D.; Liu, J.; Dong, J.; et al. The COP9 Signalosome regulates seed germination by facilitating protein degradation of RGL2 and ABI5. *PLOS Genet.* **2018**, *14*, e1007237. [CrossRef]

51. Wang, Z.; Cao, H.; Sun, Y.; Li, X.; Chen, F.; Carles, A.; Li, Y.; Ding, M.; Zhang, C.; Deng, X.; et al. Arabidopsis paired amphipathic helix proteins SNL1 and SNL2 redundantly regulate primary seed dormancy via abscisic acid–ethylene antagonism mediated by histone deacetylation. *Plant Cell* **2013**, *25*, 149–166. [CrossRef]

52. Paik, I.; Chen, F.; Ngoc Pham, V.; Zhu, L.; Kim, J.I.; Huq, E. A phyB-PIF1-SPA1 kinase regulatory complex promotes photomorphogenesis in Arabidopsis. *Nat. Commun.* **2019**, *10*, 4216. [CrossRef] [PubMed]

53. Xu, J.; Li, Y.; Wang, Y.; Liu, H.; Lei, L.; Yang, H.; Liu, G.; Ren, D. Activation of MAPK kinase 9 induces ethylene and camalexin biosynthesis and enhances sensitivity to salt stress in Arabidopsis. *J. Biol. Chem.* **2008**, *283*, 26996–27006. [CrossRef] [PubMed]

54. Hu, R.; Zhu, Y.; Shen, G.; Zhang, H. TAP46 plays a positive role in the ABSCISIC ACID INSENSITIVE5-regulated gene expression in Arabidopsis. *Plant Physiol.* **2014**, *164*, 721–734. [CrossRef] [PubMed]

55. Garcia, M.E.; Lynch, T.; Peeters, J.; Snowden, C.; Finkelstein, R. A small plant-specific protein family of ABI five binding proteins (AFPs) regulates stress response in germinating Arabidopsis seeds and seedlings. *Plant Mol. Biol.* **2008**, *67*, 643–658. [CrossRef] [PubMed]

56. Maia, J.; Dekkers, B.J.W.; Provart, N.J.; Ligterink, W.; Hilhorst, H.W.M. The re-establishment of desiccation tolerance in germinated *Arabidopsis thaliana* seeds and its associated transcriptome. *PLoS ONE* **2011**, *6*, e29123. [CrossRef]

57. Footitt, S.; Douterelo-Soler, I.; Clay, H.; Finch-Savage, W.E. Dormancy cycling in Arabidopsis seeds is controlled by seasonally distinct hormone-signaling pathways. *Proc. Natl. Acad. Sci. USA* **2011**, *108*, 20236–20241. [CrossRef]

58. Zhiguo, E.; Yuping, Z.; Tingting, L.; Lei, W.; Heming, Z. Characterization of the ubiquitin-conjugating enzyme gene family in rice and evaluation of expression profiles under abiotic stresses and hormone treatments. *PLoS ONE* **2015**, *10*, e0122621.

59. Michl-Holzinger, P.; Mortensen, S.A.; Grasser, K.D. The SSRP1 subunit of the histone chaperone FACT is required for seed dormancy in Arabidopsis. *J. Plant Physiol.* **2019**, *236*, 105–108. [CrossRef]

60. Waterworth, W.M.; Bray, C.M.; West, C.E. Seeds and the art of genome maintenance. *Front. Plant Sci.* **2019**, *10*, 706. [CrossRef]

61. Kariola, T.; Brader, G.; Helenius, E.; Li, J.; Heino, P.; Palva, E.T. Early responsive to dehydration 15, a negative regulator of abscisic acid responses in Arabidopsis. *Plant Physiol.* **2006**, *142*, 1559–1573. [CrossRef] [PubMed]

62. Toorop, P.E.; Barroco, R.M.; Engler, G.; Groot, S.P.C.; Hilhorst, H.W.M. Differentially expressed genes associated with dormancy or germination of *Arabidopsis thaliana* seeds. *Planta* **2005**, *221*, 637–647. [CrossRef] [PubMed]

63. Reddy, P.S.; Kavi Kishor, P.B.; Seiler, C.; Kuhlmann, M.; Eschen-Lippold, L.; Lee, J.; Reddy, M.K.; Sreenivasulu, N. Unraveling regulation of the small heat shock proteins by the heat shock factor HvHsfB2c in barley: Its implications in drought stress response and seed development. *PLoS ONE* **2014**, *9*, e89125. [CrossRef]

64. Sall, K.; Dekkers, B.J.W.; Nonogaki, M.; Katsuragawa, Y.; Koyari, R.; Hendrix, D.; Willems, L.A.J.; Bentsink, L.; Nonogaki, H. DELAY OF GERMINATION 1-LIKE 4 acts as an inducer of seed reserve accumulation. *Plant J.* **2019**, *100*, 7–19. [CrossRef] [PubMed]

65. Linkies, A.; Leubner-Metzger, G. Beyond gibberellins and abscisic acid: How ethylene and jasmonates control seed germination. *Plant Cell Rep.* **2012**, *31*, 253–270. [CrossRef] [PubMed]

66. Kępczyński, J.; Sznigir, P. Participation of GA3, ethylene, NO and HCN in germination of *Amaranthus retroflexus* L. seeds with various dormancy levels. *Acta Physiol. Plant* **2014**, *36*, 1463–1472. [CrossRef]

67. Oracz, K.; Bouteau, H.E.M.; Farrant, J.M.; Cooper, K.; Belghazi, M.; Job, C.; Job, D.; Corbineau, F.; Bailly, C. ROS production and protein oxidation as a novel mechanism for seed dormancy alleviation. *Plant J.* **2007**, *50*, 452–465. [CrossRef]

Article

Physical Dormancy Release in *Medicago truncatula* Seeds Is Related to Environmental Variations

Juan Pablo Renzi [1], Martin Duchoslav [2], Jan Brus [3], Iveta Hradilová [2], Vilém Pechanec [3], Tadeáš Václavek [4], Jitka Machalová [4], Karel Hron [4], Jerome Verdier [5] and Petr Smýkal [2,*]

[1] Instituto Nacional de Tecnología Agropecuaria, Hilario Ascasubi 8142, Argentina; renzipugni.juan@inta.gob.ar

[2] Department of Botany, Palacký University, Šlechtitelů 27, 783 71 Olomouc, Czech Republic; martin.duchoslav@upol.cz (M.D.); HradilovaI@seznam.cz (I.H.)

[3] Department of Geoinformatics, Palacký University, 17. listopadu 50, 771 46 Olomouc, Czech Republic; jan.brus@upol.cz (J.B.); vilem.pechanec@upol.cz (V.P.)

[4] Department of Mathematical Analysis and Applications of Mathematics, Palacký University, 17. listopadu 12, 771 46 Olomouc, Czech Republic; TadeasVaclavek@seznam.cz (T.V.); jitka.machalova@upol.cz (J.M.); karel.hron@upol.cz (K.H.)

[5] UMR 1345 Institut de Recherche en Horticulture et Semences, Agrocampus Ouest, INRA, Université d'Angers, SFR 4207 QUASAV, 49070 Beaucouzé, France; jerome.verdier@inrae.fr

* Correspondence: petr.smykal@upol.cz; Tel.: +420-585-634-827

Received: 30 March 2020; Accepted: 13 April 2020; Published: 14 April 2020

Abstract: Seed dormancy and timing of its release is an important developmental transition determining the survival of individuals, populations, and species in variable environments. *Medicago truncatula* was used as a model to study physical seed dormancy at the ecological and genetics level. The effect of alternating temperatures, as one of the causes releasing physical seed dormancy, was tested in 178 *M. truncatula* accessions over three years. Several coefficients of dormancy release were related to environmental variables. Dormancy varied greatly (4–100%) across accessions as well as year of experiment. We observed overall higher physical dormancy release under more alternating temperatures (35/15 °C) in comparison with less alternating ones (25/15 °C). Accessions from more arid climates released dormancy under higher experimental temperature alternations more than accessions originating from less arid environments. The plasticity of physical dormancy can probably distribute the germination through the year and act as a bet-hedging strategy in arid environments. On the other hand, a slight increase in physical dormancy was observed in accessions from environments with higher among-season temperature variation. Genome-wide association analysis identified 136 candidate genes related to secondary metabolite synthesis, hormone regulation, and modification of the cell wall. The activity of these genes might mediate seed coat permeability and, ultimately, imbibition and germination.

Keywords: association mapping; climate adaptation; germination; genomics; legumes; *Medicago*; plasticity; physical dormancy; seed dormancy

---

1. Introduction

Plant species exhibit a high ability for local adaptation and phenotypic plasticity that may contribute to their distribution range. While local adaptation is the genetically fixed advantage of a population under certain environmental conditions [1], phenotypic plasticity is the ability of a genotype to generate different phenotypes in response to variation in the environment [2,3]. This variation is created by mutation, recombination, and introgression, and by population genetics processes, such as genetic drift and natural selection, that determine its evolutionary fate. Understanding of the genetic basis of local adaptation and phenotype plasticity is relevant to climate change, crop production, conservation, and understanding of speciation. The combination of genomics and ecology enables genome-wide analysis to reveal the interaction between organisms and environment [4] and to identify genomic regions involved in adaptation [5]. On the other hand, phenotypic plasticity may allow species to grow and survive in different environments despite a restricted genetic base. Thus, phenotypic plasticity could be advantageous under variable environments, including climatic change [6], and may also increase species invasion success [7].

Timing of seed germination is one of the key steps in plant life, influencing the subsequent destiny of individuals as well as whole populations at determined area. Plants have evolved various mechanisms to control seed germination within- and among-seasons and in relation to the diversity of climates, habitats, and biotic pressures [8,9]. Three different kinds of dormancy have been described to allow optimal germination timing under specific environmental conditions [8,9]: (1) morphological dormancy (MD) refers to seeds that have an underdeveloped embryo and require time to grow; (2) physiological dormancy (PD) prevents embryo growth and seed germination until chemical changes occur, involving abscisic acid and gibberellins metabolism, among other factors; and (3) physical dormancy (PY) is caused by water-impermeable palisade cells in the seed coat. PY occurs in at least 18 Angiosperm families and is frequent in legumes [10–12].

Adaptation to the local environment operates through selection for successful germination and early plant establishment [13]. The prevention of germination of a certain proportion of seeds even under optimal conditions for germination reduces the risk of mortality in less predictable environmental conditions. It has been suggested theoretically [14] and shown empirically [15] that adaptation for dormancy is a bet-hedging strategy to magnify the evolutionary effect of "good" years and to dampen the effect of "bad" years, i.e., to buffer environmental variability [16]. In addition, species that frequently and reliably produce seed can afford riskier germination under unfavorable conditions (e.g., small rainfall events) because the consequences of failure to establish are less dire than for species that do not reliably produce seed [17]. Desert annuals that do not frequently and reliably reproduce are model organisms for the study of the bet-hedging strategy [18].

In order to germinate, specific environmental conditions need to be met to break the seed dormancy [8]. However, less is known about the factors which release the PY dormancy. Through experimental studies, it was shown that, in addition to scarification, wet or dry heat were found to be effective [8,10,19]. In addition, natural conditions, such as temperature and soil moisture oscillations, are the major players [20,21]. Laboratory studies have demonstrated an association between seed responsiveness to temperature and environmental thermic characteristics [10,22,23]. However, only limited data are available on how and why PY varies inter- and intra-specifically in natural ecosystems [24]. Legumes are thus a model example for studies of PY dormancy patterns in relation to environmental variations. The study of Rubio de Casas et al. [25] showed a latitudinal gradient in PY dormancy in legumes. Thus, PY dormancy increases from regions with long growing seasons (e.g., tropical climate) in lower latitudes to regions with a seasonal climate in higher latitudes. However, there are some studies of intraspecific PY dormancy variation along environmental gradients in several legume species [10,26–29] that are in disagreement with the results of Rubio de Casas et al. [25]. *Medicago truncatula* (barrel medic) is an annual, diploid, self-fertile species with a natural geographic distribution across the Mediterranean Basin. Phenotypic variation among populations has been explained by the adaptation to local environmental conditions [30]. *M. truncatula* offers an excellent model to study seed dormancy in relation to genetic and environmental factors because within its range it inhabits environments with rather contrasting climatic conditions, differing not only in mean annual temperature and precipitation, but also in within- and across-season variability (unpredictability). Its seeds exhibit both physical and physiological dormancy. Physiological dormancy in *M. truncatula* seeds is non-deep, and is removed during the seed ripening period [31,32]. The short after-ripening period to overcome PD (<3 months) determines that PY release is the most important trait to regulate the timing of seedling emergence. Despite this, most germination studies in *M. truncatula* eliminate the influence of PY dormancy through prolonged periods of storage (>9 months) and/or by seed scarification [33,34]. Large georeferenced collections, a reference genome, and a high-density single nucleotide polymorphism (SNP) map of more than 260 genotypes of *M. truncatula* are available [30,35] and were used for study of the association between the genome and the environment in relation to flowering [35–37]. We have taken advantage of this georeferenced collection to analyze the patterns of dormancy release in 178 accessions of *M. truncatula* originating from various environments in the Mediterranean basis and tested seeds under alternating temperatures. The following questions were addressed: (i) Is there variation in physical seed dormancy among accessions to temperature treatments? (ii) Which of the ecological factors acting as potential adaptation drivers are correlated with dormancy? (iii) Are there any candidate genes that might be related to seed dormancy release in *Medicago*, using genome-wide association (GWAS) analysis?

Our study showed that phenotypic plasticity of final dormancy was significantly correlated with increased aridity, suggesting that plastic responses to external stimuli provide seeds with strong bet-hedging capacity and the potential to cope with high levels of environmental heterogeneity. Genome-wide association analysis performed on seven seed dormancy traits and three bioclimatic variables identified 136 candidate genes as potential regulators of physical dormancy. A large proportion of candidate genes were annotated as involved in synthesis of secondary metabolites, in cell wall modification, and hormone regulation. The knowledge about the regulation of seed dormancy by environmental factors could be extended to other legume species, particularly to crop wild relatives of economically important crops, such as chickpea, lentil, and faba bean. In addition, it can be used in a conservation biology context for the management of endangered plant species in relation to climate change.

2. Results

2.1. Responses of Dormancy Traits of Medicago Accessions to Experimental Temperature Treatments

Most dormancy traits exhibited a near normal distribution, and a wide range of variability (Figure 1A; Supplementary data Figure S1). Final PY dormancy (FPYD), a proportion of dormant seeds after 88 days of incubation onto water-saturates, ranged from 34% to 100% with mean 80% (SD = 15) at 25/15 °C treatment, and from 4% to 94% with mean 60% (SD = 19) at 35/15 °C treatment. Comparison of responses of each accession to two temperature treatments showed a remarkable effect of larger temperature alternation on dormancy release in a majority of accessions (Figure 1B). The germination pattern (area under curve, AUC_M) ranged from 3 to 79, and, similarly to FPYD, larger temperature alternation increased the dormancy release (AUC_{25}: mean ± SD 24 ± 13, range 0–79; AUC_{35}: 34 ± 17, range 5–82), except for some accessions (16%) where the differential (AUC_{35-25}) was negative (Figure 1A and Figure S1). Both phenotypic plasticity indexes, based on the minimum and the maximum value among the two temperature treatments divided by the maximum value (PI), showed large ranges with mean PI_{AUC} being slightly higher (0.56 ± 0.29, range 0.00–1.00) than mean PI_{PY} (0.43 ± 0.23, range 0.00–0.91) (Figure 1A and Figure S1). All dormancy traits were moderately to strongly correlated (up to |0.75|, excluding $FPYD_M$ and AUC_M with some correlations up to |0.94|), except for PI_{AUC}, which was significantly correlated only with AUC_{35} and AUC_{25} (Figure 1A; Supplementary data Figure S2, S3).

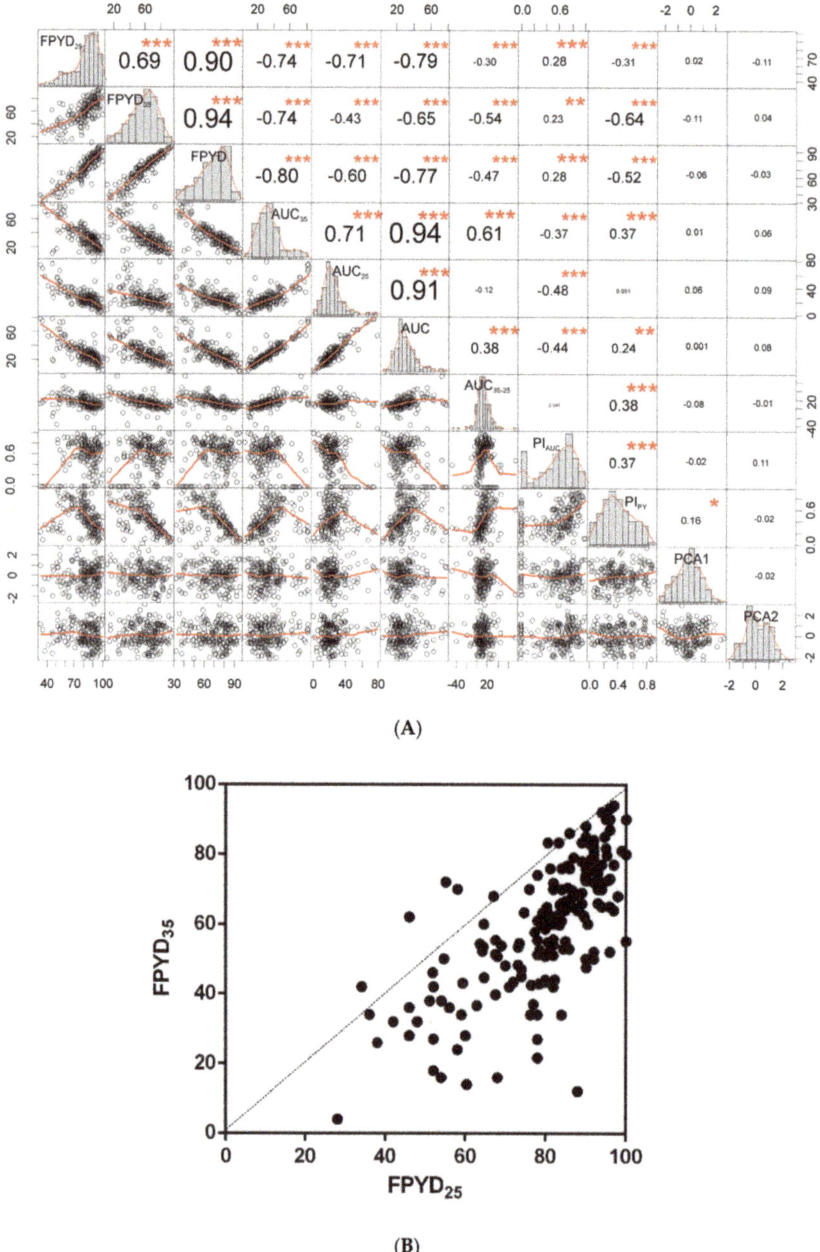

(A)

(B)

Figure 1. Correlations among dormancy release traits. (**A**) Correlation chart of dormancy release traits and ordination scores of environmental principal component analysis, PCA (first two ordination axes; PCA1, PCA2; see Figure 2A,B). The distribution of each variable is shown on the diagonal. Bellow the diagonal the bivariate scatter plots with a fitted smooth line (loess) are displayed. Above the diagonal the value of the Pearson correlation coefficient plus the significance level as stars are displayed (* $p \leq 0.05$, ** $p \leq 0.01$, *** $p \leq 0.001$). (**B**) Relationship between final physical dormancy (PY) dormancy of each accession under two temperature treatments (FPYD$_{35}$ and FPYD$_{25}$).

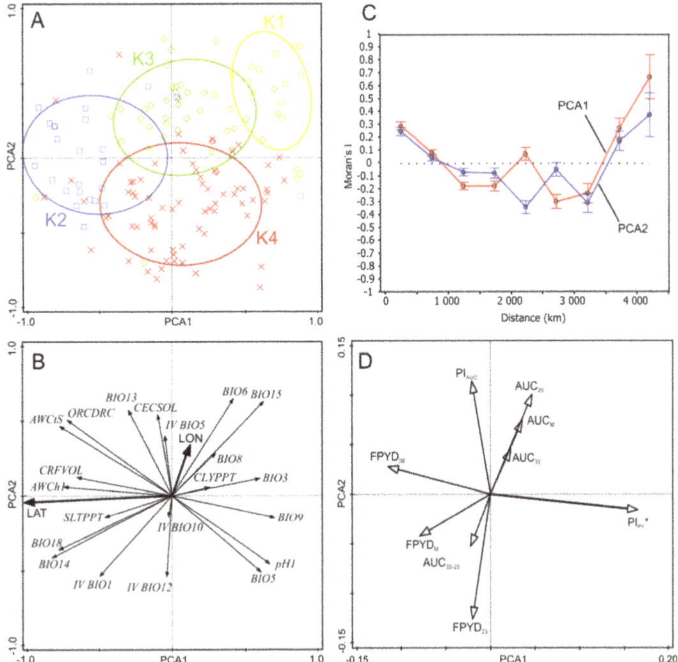

Figure 2. Principal component analysis (PCA) of selected bioclimatic and soil variables of *Medicago* accessions and multiple correlations of dormancy traits with ordination axes. (**A**,**B**) Principal component analysis (PCA) of selected bioclimatic and soil variables of Medicago accessions. Each accession is classified according to cluster analysis of environmental variables into one of four clusters (see Methods). The ellipses were created based on a model of bivariate normal distribution of the cluster class symbols (estimated from a variance–covariance matrix of their X and Y coordinates) to cover 95% of that distribution' cases. A comparison of selected environmental variables among clusters is shown in Supplementary fileSupplementary file Tables S1 and S2. Vectors of geographic variables (latitude, longitude) were added into the diagram after PCA to visualize spatial gradients of environment. Variables BIO14 and 18 were log(x+1) transformed before analyses. (**C**) Spatial autocorrelation diagram of Moran's I for the first two ordination axes of PCA (PCA1, PCA2). Mean ± 95% CI of I for respective distance class is calculated. (**D**) Multiple correlations of dormancy traits with the first and the second ordination axes of the environmental PCA. Each arrow points in the direction of the steepest increase of the values for corresponding dormancy trait. The angle between arrows indicates the sign of the correlation between the variables. The length of the variable arrows is the multiple correlation of that variable with the ordination axes. Dormancy trait significantly correlated ($p \leq 0.05$, spatial correlation) with any ordination axis has an asterisk.

2.2. Associations of Environmental Gradients with Dormancy Traits of Medicago Accessions

Principal component analysis (PCA) of a reduced data set containing 14 climatic and eight soil variables revealed two clear environmental gradients (Figure 2A,B). The first ordination axis explained 30.4% of the total variation and can be interpreted as the gradient of aridity that is tightly correlated with latitude (i.e., the north–south gradient). Climatic variables with the highest positive/negative correlation with the fist ordination axis represent temperatures of the warmest month (BIO5; Pearson's r = 0.61 ***) and the driest quarter (BIO9, r = 0.71 ***), isothermality (BIO3, r = 0.59 ***), precipitation of the driest month (BIO14; r = −0.83 ***) and precipitation of the warmest quarter (BIO18, r = −0.78 ***). Concerning soil variables, pH index is positively correlated (r = −0.69 ***) while soil organic carbon content (ORCDRC, r = −0.75 ***), available soil water capacity (AWCh1, r = −0.75 ***), and saturated water content (AWCtS, r = −0.79 ***) are negatively correlated with the first axis. Latitude (r = −0.77 ***) but not longitude (r = 0.07) is strongly negatively correlated with the first axis. The second ordination axis explained 17.3% of the total variation and can be interpreted as combined gradient of seasonality and inter-annual variability, with weak geographic (i.e., west–east) trend (latitude: r = −0.01, longitude: r = 0.24 ***). The most correlated variables with the second axis were precipitation seasonality (BIO15, r = 0.61 ***) and minimal temperature of the coldest month (BIO6, r = 0.63 ***), but inter-annual variability of temperature (IV BIO1, r = −0.54 ***) and precipitation (IV BIO12, r = −0.54 ***) also had high correlation coefficients. Both synthetic environmental variables (PC1, PC2) were spatially structured as revealed by Moran's I correlogram (both $p < 0.001$), showing positive autocorrelation at short and large distance classes and negative autocorrelation at intermediate distance classes (Figure 2C). Inspection of dormancy trait correlations with ordination axes representing synthetic environmental variables showed that only PI_{PY} was significantly correlated with the first ordination axis (r = 0.16 *), even after correction for spatial autocorrelation ($p = 0.032$). Other dormancy traits did not show any significant correlation with the first two ordination axes of PCA (Figure 2D; Supplementary data Figure S2). Neither dormancy trait showed any spatial autocorrelation (all Moran's I correlograms had $p > 0.40$, not shown).

Separate analyses of relationships between each dormancy trait and each bioclimatic and soil variable showed that only one dormancy trait (PI_{PY}) was significantly correlated with more environmental variables, while other dormancy traits were either not correlated or showed weak correlations with some environmental variables (Supplementary data Figures S2, S3 and S4, Table S4). Specifically, PI_{PY} was clearly related to the gradient of aridity, i.e., PI_{PY} increases with increasing temperatures and decreasing precipitation and decreasing available soil water capacity (Figure 2; Supplementary data Figure S2). However, there were three climatic variables, i.e., IV BIO1, IV BIO5, and IV BIO10, which showed significant correlations with a majority of dormancy traits (Supplementary data Table S5). Specifically, final PY dormancy ($FPYD_M$, $FPYD_{25}$, $FPYD_{35}$) slightly increased with increasing inter-annual variation in temperatures of the warmest quarter (all r = ~0.19 *).

Four macro-environmental groups of *Medicago* accessions (Figures 2A and 3, Supplementary data Table S3) differed in slopes of the FPYD across two experimental temperature treatments (Figure 4). Considering each experimental year separately, accessions from arid conditions (clusters K1 and K4, Supplementary data Table S3) consistently showed higher FPYD at 25/15 °C and lower at 35/15 °C. In contrast, FPYD of accessions from K2 (less arid conditions) did not change significantly in response to different temperature treatments (Figure 4).

2.3. Association Analysis of Dormancy Traits

In order to identify molecular mechanisms underlying physical dormancy and its adaptability, we performed genome-wide association analyses for all dormancy traits ($FPYD_{25}$, $FPYD_{35}$, AUC_{25}, AUC_{35}, AUC_{35-25}, PI_{PY}, PI_{AUC}) and three bioclimatic variables (BIO1, BIO9, BIO12) on 178 accessions. Corresponding Manhattan plots for these analyses are provided in Supplementary data Figure S5. Quantile–quantile (Q-Q) plots confirmed that FarmCPU was a more suitable model to perform association studies (Supplementary data Figure S6). Most significant Quantitative Trait Nucleotides (QTNs) were identified with AUC_{25}, AUC_{35-25}, $FPYD_{25}$, PI_{AUC} and all three bioclimatic variables. To provide a list of significant QTNs, we defined a threshold of 10^{-7} (except for PI_{PY}, where we used a threshold of 10^{-4}). 136 candidate genes were identified as potential regulators of physical dormancy (Supplementary data Table S6). A large proportion of candidate genes was annotated as involved in synthesis of secondary metabolites, in cell wall modification, and hormone regulation. We performed an over-representation analysis with these 136 candidate genes using a hypergeometric test with Bonferroni correction and this revealed three biological functional Gene Ontology (GO) classes statistically overrepresented (Supplementary data Table S7) and acting as potential regulators of dormancy: response to oxidative stress (GO:0006979), oxidation reduction (GO:0055114), and response to chemical stimulus (GO:0042221). Candidate genes belonging to these three GO classes are indicated in Table 1.

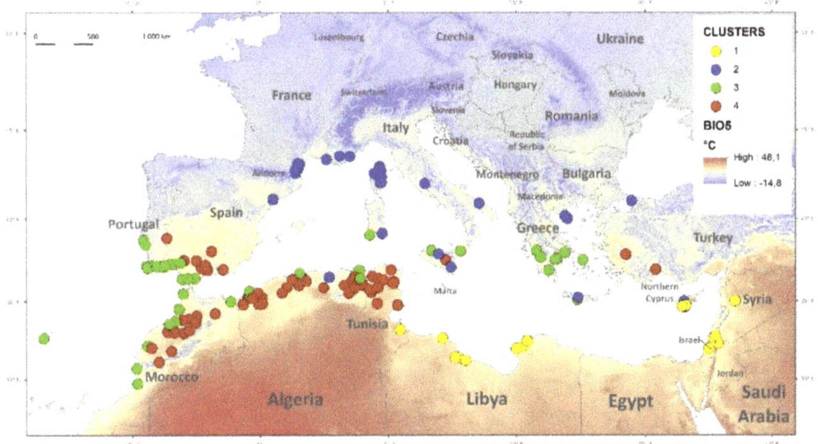

Figure 3. Geographic distribution of studied *Medicago truncatula* accessions classified into four clusters based on climatic and soil conditions, using Ward's minimum-variance linkage of Euclidean distance. Grey dots indicate K1, green K2, light blue K3, and yellow K4 cluster, placed on the background of BIO5 (precipitation in the wet quarter).

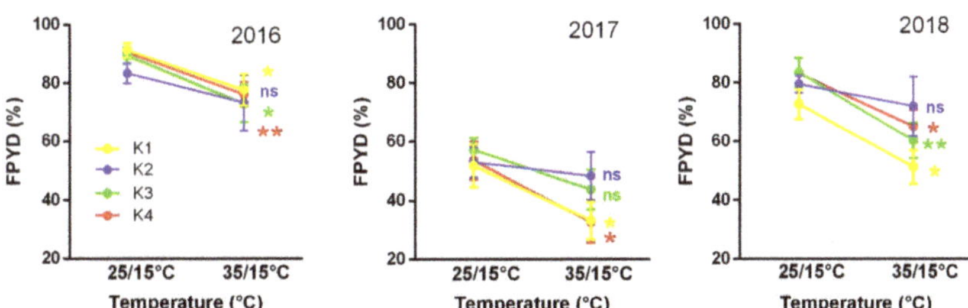

Figure 4. Reaction norms to changes in experimental temperature (25/15 °C, 35/15 °C treatments) on final PY dormancy of seeds for K1–K4 macro-environmental clusters in three experimental years (2016, 2017, and 2018). Vertical bars indicate ± SE. Asterisk (* $p \leq 0.05$ and ** $p \leq 0.01$) indicates significant differences between temperatures for each cluster.

Table 1. List of Quantitative Trait Nucleotides (QTNs) identified by genome-wide association (GWA) analysis of each dormancy trait and belonging to one of three biological function over-represented in our complete list of candidate QTNs. Corresponding chromosome locations and p-values of QTNs are indicated as well as the closest gene ID within +/- 10 kb genomic interval and its corresponding annotations (Mtv5 annotations, Mtv4 annotations, gene ontology, and gene description). For complete list and description of all identified QTNs see Supplementary data Tables S6 and S7.

Chrom	Position of QTN	P-value	Candidate gene ID within +/- 10kb interval (Mtv5)	Gene ID v4	Gene annotation	Gene description
1	34656783	4.25×10^{-11}	MtrunA17Chr1g0184131	Medtr1g070110	Hyoscyamine 6-dioxygenase	secondary metabolism.flavonoids.dihydroflavonols
1	49585766	1.36×10^{-7}	MtrunA17Chr1g0204011	Medtr1g101830	Peroxidase	misc.peroxidases
1	50761951	2.79×10^{-7}	MtrunA17Chr1g0205571	Medtr1g104590	Primary-amine oxidase	misc.oxidases - copper, flavone etc
1	50761951	4.23×10^{-8}	MtrunA17Chr1g0205541	Medtr1g104550	Primary-amine oxidase	misc.oxidases - copper, flavone etc
2	10895258	5.60×10^{-9}	MtrunA17Chr2g0291751	Medtr2g028980	Peroxidase	misc.peroxidases
2	12184616	5.76×10^{-9}	MtrunA17Chr2g0293411	Medtr2g031920	Ent-kaurenoic acid oxidase 2	hormone metabolism.gibberelin.synthesis-degradation.ent-kaurenoic acid hydroxylase/oxygenase
4	53502844	5.22×10^{-8}	MtrunA17Chr4g0061311	Medtr4g109360	UDP-glucose 6-dehydrogenase	cell wall.precursor synthesis.UDP-Glc dehydrogenase (UGD)
4	53502844	3.62×10^{-8}	MtrunA17Chr4g0061381	Medtr4g109470	Flavonoid 3′-monooxygenase	secondary metabolism.flavonoids.dihydroflavonols.flavonoid 3-monooxygenase
4	61245281	2.56×10^{-7}	MtrunA17Chr4g0072091	Medtr4g127670	Peroxidase	misc.peroxidases
5	4768980	2.23×10^{-12}	MtrunA17Chr5g0400421	Medtr5g014250	Beta-amyrin 11-oxidase-like	hormone metabolism.gibberelin.synthesis-degradation.ent-kaurenoic acid hydroxylase/oxygenase
5	33045218	2.41×10^{-7}	MtrunA17Chr5g0432331	Medtr5g074710	Peroxidase	misc.peroxidases
5	33045218	2.41×10^{-7}	MtrunA17Chr5g0432361	Medtr5g074770	Peroxidase	misc.peroxidases
6	28332981	4.21×10^{-10}	MtrunA17Chr6g0474391	Medtr6g464620	Gibberellin 3-beta-dioxygenase	hormone metabolism.gibberelin.synthesis-degradation.GA20 oxidase
6	34330903	7.87×10^{-6}	MtrunA17Chr6g0479681	Medtr6g072490	Cytokinin hydroxylase-like	misc.cytochrome P450
8	6442861	1.76×10^{-9}	MtrunA17Chr8g0343001	Medtr8g018650	Seed linoleate 9S-lipoxygenase	hormone metabolism.jasmonate.synthesis-degradation.lipoxygenase
8	48706047	1.32×10^{-9}	MtrunA17Chr8g0391921	Medtr8g105630	Glutathione peroxidase	redox.ascorbate and glutathione.glutathione

3. Discussion

Medicago truncatula is a representative of species adapted to the typical Mediterranean climate characterized by strong seasonality with hot and dry summers often with large diurnal temperature oscillations followed by main rainfall in the autumn. In particular, the eastern and southern zones of the Mediterranean-desert transitions are associated with increased aridity [38]. Summer drought limits growth and is the major cause of seedling mortality, while winter cold limits vegetative growth. High oscillating temperatures are hypothesized [8] to be one of the main triggers of PY dormancy release and it was confirmed experimentally in several legume species, including *Lupinus, Trifolium, Pisum* [10,26–28] and, in this study, *Medicago*. However, these studies tested only the effect of temperature variation, while more factors vary in nature [39]. In particular, soil moisture oscillation is very difficult, if not impossible, to mimic in laboratory conditions. The effect of water potential on seed germination was tested only as a static component [40,41]. As a result, our experimental setup could reveal only temperature-related dormancy release.

3.1. Association between Seed Dormancy Traits and the Environment

Variation in germination strategy is particularly relevant for plants inhabiting unpredictable environments and is consistent with seed function, and securing the next generation in time and space [39]. In our study, seed dormancy release varied among accessions and years, and this could potentially act as a mechanism that favors the persistence of the seed in the soil and helps to distribute genetic diversity through time [24,42]. However, when taking average estimates of various traits characterizing absolute dormancy release (i.e., FPYD, AUC, and AUC$_{35-25}$), no clear relationships were found between synthetic environmental (macroclimatic) clines (expressed as PCA axes) and mean dormancy status per accession (Figure 2). Our observations thus seem to be in agreement with study of Mediterranean wild lupines [29] or perennial woody legume (*Vachellia aroma*) along a precipitation gradient that did not find any relationship between climate and dormancy release [43]. In contrast, *Arabidopsis thaliana* [44,45] and other winter-annual species, such as *Betta vulgaris* subsp. *maritima, Biscutella didyma, Bromus fasciculatus*, and *Pisum sativum* subsp. *Elatius*, showed a cline in dormancy [10,43,46,47]. In particular, more dormant genotypes of mentioned taxa occurred in lower latitudes or more arid habitats with seasonally unpredictable precipitation and less dormant in higher latitudes or more humid habitats, suggesting dormancy is an adaptation securing population survival in less predictable conditions. However, in arid and unpredictable environments, there are also species called "risk-taking" with lower dormancy and rapid germination in response to lower rainfall events. The high and reliable seed production determines that the consequences of failure to establish in these species are less dire [17].

Does the absence of a relationship between average estimates of dormancy release in *Medicago* and macroclimatic clines within our dataset suggest that selection does not operate on dormancy traits? Taking into account plasticity of dormancy release between temperature treatments revealed that physical dormancy plasticity (PI$_{PY}$) increases with increasing aridity (2D, Figures 3 and 4; Supplementary data Figure S3). It follows, therefore, that more plastic behavior can potentially distribute germination across the year and act as a within-season bet-hedging strategy [48], suggesting that in more unpredictable environments genetic components of phenotypic variance may be lower and thus a reduced evolutionary response to selection would be possible [24]. The bet-hedging strategy is thus positively associated with more arid habitats as found in pea [10], and plastic responses provide potential to cope with high levels of environmental heterogeneity [49]. On the other hand, the responsiveness of the accessions of the macro-environmental clusters K1, K3 (except in 2017), and K4 to high temperatures (35/15 °C) in relation to cluster K2 (Figures 2–4) could be one of the main triggers of PY dormancy release under an arid and unpredictable environment. Occasional precipitation and hot temperatures in summer would accelerate the PY dormancy release and germination after overcoming the PD dormancy. Earlier emergence can profit from a long growing season, providing a competitive and reproductive advantage for limited resources. Recently, ten Brink et al. [50] showed that more unpredictable natural environments can select earlier within-season germination phenology. In addition, we found an increase, although slight, in PY dormancy for accessions originating from sites with higher among-season temperature variations expressed by IV indexes. This could be an among-season bet-hedging strategy, which increases PY dormancy under more unstable summer temperatures between years, and might thus contribute to avoidance of increased germination under adverse temperatures condition or "false breaks" (seed germination outside the optimal growing season) [51].

3.2. Potential Shortcomings of the Study

Although dormancy is genetically determined, it also depends on the environmental conditions experienced by the mother plant and the subsequent status of the seed [24,39]. This was shown in several taxa, including *Trifolium* and *Medicago* [24,27,28,52,53]. There are several possible sources of variability, from the effect of the maternal plant status (drought, photoperiod, nutrition) to natural variation within a population or even the same individual [54]. Distinguishing between maternal and environmental effects is difficult. The information on the maternal environmental can be mediated via the nutrition, phytohormones, or gene expression levels and variability in the seed sensitivity [39]. All this might contribute to *M. truncatula* seed stock generated in our study. To minimize environmental maternal effects, we grew the accessions under common garden conditions (glasshouse), but these were to some extent variable between years. In 2016, the accessions were sown February to April and harvested in a hot July with a day temperature over 35 °C, while in 2017 and 2018, they were sown in September and grew over winter, flowered in February, and matured in April, with day temperatures in the glasshouse around 28 °C. The higher temperature in 2016 during the seed filling period resulted in more dormant seeds in some accessions in relation to 2017 and 2018 (Supplementary data Figure S8). In addition to this, seeds from different accessions differ up to 3 weeks in maturation due to differences in flowering time. Moreover, the individual seed stock from a given accession was harvested during a period of about 3 weeks, which could be possibly synchronized by different sowing. This needs to be considered in follow up studies.

Our analysis was also likely impacted by several factors inherent to the available *Medicago* set. At first, there is substantial imprecision in the GPS localization of the origin of some accessions [55] leading consequently to incorrectly extracted environmental factors [56–58]. Secondly, there is geographical bias towards the western part of the Mediterranean with underrepresented parts of the native species range, such as Italy, Adriatic Sea coast, Turkey, Lebanon, and Israel. In addition, the characteristics of WorldClim data (averages in terms of time and also space) mask the micro-ecological pattern and, in geographically complex regions, environmental conditions change considerably over short spatial scales, such that neighboring populations can be subject to different selective pressures, as found in the study of seed dormancy of Swedish *A. thaliana* accessions [59].

3.3. Genetic Basis of Seed Dormancy Release in Legumes

In contrast to the seed development, the genetic basis of *Medicago* seed germination was studied only by Dias et al. [60]. These authors, however, focused on true germination, e.g., radicle emergence, removing the seed coat prior to testing and thus assessing physiological dormancy, while we were interested in PY dormancy executed by seed coat permeability. Since the dormancy is indirectly evaluated as the release from dormancy by imbibition and germination, the detected candidate genes might be related to the changes in the seed coat mediated imbibition process rather than dormancy per se. There was no overlap in associated loci between the tested seed dormancy traits, despite the detection of a similar set of candidate genes (Supplementary data Table S7). This is similar to other tested quantitative and complex traits, such as drought and biomass, where different traits had different candidate genes [61]. We have detected four genes active in flavonoid and phenylpropanoid biosynthesis pathways leading either to flavonoids or via polymerization to lignins impregnating the seed coat [12,62]. Notably, homologue genes were detected when comparing dormant and non-dormant pea seed coat expression [63]. Furthermore, hydrolytic enzymes such as xyloglucan 6-xylosyltransferase, xylogalacturonan β-1,3-xylosyltransferase involved in plant cell wall modification were identified. Notably, one of the Quantitative Trait Loci (QTL) identified in biparental mapping of *Medicago* seed germination also encodes xyloglucan endotransglucosylase [60]. The β-1,3-Glucanase (EC 3.2.1.39) plays roles in the regulation of seed germination, dormancy, and in the defense against pathogens. The β-1,3-glucan layer is in the seed coat of cucurbitaceous species and confers seed semi-permeability [64]. In tobacco seeds, β-1,3-Glucanase was shown to be at the micropylar part of the endosperm prior to radicle protrusion, and seems to play a role in cell wall loosening [65]. Pectinesterases were isolated from germinating seeds of various species and are assumed to play an important role in loosening cell walls [66]. Polygalacturonases (EC 3.2.1.15) are another cell wall degrading enzyme. These were shown to play an essential role in pollen maturation and in pectin metabolism during fruit softening and weakening of the endosperm cell walls [67]. Exo-(1→4)-β-galactanases (EC 3.2.1.23) play various roles in physiological events, including cell wall expansion and degradation during soft fruit ripening, and were found to be involved in the mobilization of polysaccharides from the cotyledon cell walls of *Lupinus angustifolius* following germination [68]. Nitric oxide (NO) was recently shown to be involved in plant development including seed germination [69]. NO-dependent protein post-translational modifications are proposed as a key mechanism underlying NO signaling during early seed germination. Our GWAS analysis identified seven putative peroxidase and thio-/peroxiredoxin genes. Peroxiredoxins (EC 1.11.1.15) catalyze the reduction of hydroperoxides, conferring resistance to oxidative stress. Recent studies have demonstrated that Reactive Oxygen Species (ROS) have key roles in the release of seed dormancy, as well as in protection from pathogens [70–72]. Thioredoxins were identified to promote seed germination [73]. Peroxidases (EC 1.11.1.7) are also implicated in lignin/suberin formation during the polymerization of monolignols synthetized in the final steps of the phenylpropanoid pathway [74]. Phytohormones and especially gibberellins are known to play important roles in seed development and germination (reviewed in [75]), and since *Medicago* seeds have both physical and physiological dormancy, it is not surprising to find gibberellin 20-oxidase and two ent-kaurenoic acid oxidase genes to be associated with dormancy release (AUC_{25}) or environmental factors (BIO12), respectively. The genomic signature of *M. truncatula* adaptation to the climate was studied by Yoder et al. [76] using essentially the same set of lines. They analyzed the relationship to BIO1, BIO3, and BIO16; thus, there is only overlap in BIO1 (annual mean temperature) with our study. The different candidate genes were detected. This could be attributed to differences in accessions and analytical methods, as well as *Medicago* genome versions. However, despite these differences, some similar genes were detected, such as 1, 3-glucanase or kinases. Several kinases and disease resistance (TIR-NBS-LRR class) genes are associated with dormancy release traits. These have been implicated in pathogen sensing and host resistance, which might reflect the sensing of cell wall modulating enzyme activities, similar to pathogen attack [77]. As the result of bet-hedging, seeds in the soil form long-term seed banks where they need to be protected from microbial decay by presence of secondary metabolites, as well as seed defense enzymes [72,77]. Therefore, one of the possible future directions of seed dormancy release studies should be the study of seed–soil–microbiome relationships and seed coat enzymatic activities.

4. Materials and Methods

4.1. Plant Material

Seeds of *Medicago truncatula* inbred lines were selected from HapMap collection [36,76] based on accuracy of coordinates and were obtained from INRA, Montpellier, France and University of Minnesota, USA. Plants were grown in glasshouse conditions at the Department of Botany, Palacký University, Olomouc, Czechia, from March to July (2016) and from September to May (2017, 2018). Plants were cultivated in 3 L pots with sand peat substrate (1:9) mixture (Florcom Profi, BB Com Ltd., Letohrad, Czechia), watered as required, and fertilized weekly (Kristalon Plod a Květ, Agro, Czechia).

Temperature varied according to weather from a minimum of 15 °C during winter to a maximum of 40 °C in late spring. Supplementary light was provided (Sylvania Grolux 600 W, Hortilux Schreder, Holland) to extend the photoperiod to 8 h during September–February and to 14 h from February to stimulate flowering. Mature pods were collected, packed in paper bags, and dried at 20 °C and 60–63% relative humidity to allow post ripening for a period of 4 to 6 weeks prior to testing. Sufficient seed stock was obtained from 178 accessions using equipment made in-house.

4.2. Seed Dormancy Release Experiments

Release of seed dormancy was tested as imbibition (e.g., uptake of water) and terminated when the radicle protruded the seed coat [10]. As our study was aimed to study PY release and not the germination, the values we used in all subsequent analysis correspond to imbibed seeds. To mimic natural conditions, two temperature treatments (alternating temperatures of 35/15 °C and 25/15 °C at 14/10 h (day/night) regime) were applied to intact seed batches (50 seeds, in 2 to 3 replicas per treatment). Seeds were placed onto water saturated filter papers (Whatman Grade 1, Sigma, CZ) in 60 mm Petri dishes (P-Lab, CZ) in temperature-controlled chambers (Laboratory Incubator ST4, BioTech, CZ). In order to prevent fungal growth (as seed sterilization would alter seed coat properties), fungicide (Maxim XL 035 FS; containing metalaxyl 10 g and fludioxonil 25 g) was applied. Seeds were monitored at 24 h intervals for total of 88 days. After each scoring, the plates were randomly relocated within chambers. Germinated seeds (e.g., when the radicle protruded from the testa) were removed. At the end of the testing, remaining seeds were scarified and let germinate to verify their viability. This typically was over 98%. Although we selected macroscopically intact seeds for experiments, we cannot exclude some microscopic cracks on seeds resulting from mechanical damage during threshing. We observed that in the course of the first hours of seed germination experiments, a certain proportion of the seeds imbibe. Therefore, we subtracted the first day imbibition value from the analysis. In 2016, 2017, and 2018, a total of 147, 74, and 130 accessions were tested (Supplementary data Figure S8). In total, seeds of 178 accessions were included in the experiments (Supplementary data Table S1). Forty-seven accessions were tested in all three years, and 129 accessions in at least two years.

4.3. Evaluation of Dormancy and Germination Traits

Several statistics (traits) characterizing dynamics and final state of dormancy release of seeds of each accession for each treatment (i.e., 25/15 °C and 35/15 °C) were calculated as follows: *(i)* Final PY dormancy (%; $FPYD_{25}$, $FPYD_{35}$): represents percentage of dormant seeds at the end of experiment after excluding seeds germinating during the first day of each experiment (i.e., 100 – final germination percentage after 88 days + germination percentage after first day), calculated separately for two germination treatments. *(ii)* Germination pattern (AUC_{25}, AUC_{35}): this trait represents the area under curve (AUC) coefficient that takes into account both dynamics of germination as well as final germination percentage. Original germination data (daily counts of imbibed seeds) were considered as discrete realizations of an asymptotically continuous process, approximated by spline functions [78,79]. The resulting smoothing spline, called the absolute germination distribution function (AGDF; as applied in pea [10]), was used in analysis. Accordingly, the area under curve (AUC) of the spline function takes into account both the course of the germination as well as the final score of germinated seeds, which captures the dynamics of seed germination better than existing germination coefficients [80,81]. High AUC values mean rapid and early germination of the majority of seeds. *(iii)* $FPYD_M$ and AUC_M: these two coefficients represent means of respective coefficients estimated for the two temperature treatments (e.g., $FPYD_M = (FPYD_{25} + FPYD_{35})/2$)). *(iv)* Germination response (AUC_{35-25}): this is the difference between two AUC (i.e., $AUC_{35} - AUC_{25}$) of each accession calculated for two germination treatments (i.e., 35/15 °C and 25/15 °C). Higher absolute values of germination response mean larger differences in germination pattern of the same accession between germination treatments, while the sign of the difference suggests which of the treatment shows the larger AUC. *(v)* Phenotypic plasticity index of the germination pattern (PI_{AUC}) and *(vi)* Phenotypic plasticity index of final PY dormancy (PI_{PY}): these traits were calculated for each accession as (trait$_{max}$ − trait$_{min}$)/trait$_{max}$, where trait$_{max}$ and trait$_{min}$ were, respectively, the maximal and minimal value of the trait measured on the same accession across the two temperature treatments (25/15 °C and 35/15 °C). These estimates characterize the maximal plastic capacity of an individual in variable environments without taking into account the direction of the plastic response or the change in intensity with environmental variation. Phenotypic plasticity index ranges from 0 (no plasticity) to 1 (maximal plasticity) [2]. For the purpose of multivariate analyses, we calculated the average of each dormancy trait for each accession over the years. Consequently, the matrix of averages of each dormancy trait for each accession was used in multivariate analyses. Multicollinearity among variables was assessed by the variance inflation factor (VIF) for quantitative traits using the library usdm in R [82]. Only variables whose VIF was lower than 15 were retained in the analyses. Except for FPYD$_M$ and AUC$_M$, none of the above-mentioned traits had a collinearity problem.

4.4. Extraction of Environmental Variables and Spatial Accuracy

Due to different spatial accuracies of accessions and in order to minimize the spatial error caused by imprecise coordinates, we developed a geoprocessing model in the ArcGIS PRO environment [83]. The model automated the calculation of mean values of selected variables from within a 5 km buffer around each collection site in order to smooth the uncertainty caused by imprecise localization. The WorldClim database version 2.0 [84] was used to extract climatic data (period 1970–2000) from GeoTIFF rasters in the WGS-84 coordinate system (EPSG: 4326) with a spatial resolution of 30 arc-seconds (~1 km). Bioclimatic variables (BIO1–BIO19) were derived from the monthly temperature and rainfall values [85], and represent annual trends (e.g., mean annual temperature BIO1, annual precipitation BIO12), seasonality (e.g., annual range in temperature and precipitation BIO4 and BIO15) and extreme or limiting environmental factors (e.g., the temperature of the coldest and warmest month BIO5 and BIO6, and amount of precipitation in the wet and dry quarters BIO16 and BIO17). In order to determine the inter-annual variability in selected bioclimatic variables (BIO1, BIO5, BIO10, and BIO12) during the period 1981–2010, the index of variability (IV) was calculated following the percentile-analysis method [86]. To obtain yearly mean values for years 1981–2010, we used 2 m air temperature (Kelvin degrees) and 2 m specific humidity (kg of water/kg of air) hourly data from the Modern Era Retrospective Analysis for Research and Applications Reanalysis (MERRA) 2D Incremental Analysis Update atmospheric single-level diagnostics product (MAT1NXSLV), provided by the NASA Global Modelling and Assimilation Office [87]. Data were interpolated in spatial resolution of 2.5 arc-minutes. Temperature data were converted to degrees of Celsius (BIO1, 5, and 10). The resulting BIO12 (1981–2010) describes the annual mean of specific humidity instead of cumulative annual rainfall.

Twelve month means (BIO1, 5, 10, and 12) for each site were calculated as follows:

$$IV = [(90\text{th percentile} - 10\text{th percentile})/50\text{th percentile}] * 10 \tag{1}$$

The different IV classes are low (IV <0.50), low–moderate (0.50–0.75), moderate (0.75–1.00), moderate–high (1.00–1.25), high (1.25–1.50), very high (1.50–2.00), and extreme (IV >2.00).

Soil data were extracted from the SoilGrids database [88]. SoilGrids prediction models are fitted using over 230,000 soil profile observations from the World Soil Information Service (WoSIS)database and a series of environmental covariates. Covariates were selected from environmental layers from Earth observation-derived products and other environmental information, including climate, land cover and terrain morphology [89].

4.5. Climate and Soil Characteristics of Localities of Studied Accessions

The set of accessions originates from rather contrasting climatic conditions (Supplementary data Table S2). Mean annual temperature ranges from ca 9 to 22 °C and annual precipitation ranges from 154 to 1028 mm; consequently, temperature annual range (min–max) is from 13 to 35 °C. Some accessions originate from sites with minimal winter temperatures below zero while maximal temperature of warmest months is rather similar among accessions. However, sites differ considerably in precipitations of driest and warmest periods. The basic descriptive statistics of the index of variability (IV) for BIO1, BIO5, BIO10, and BIO 12 are present in Table S2. IV BIO1 ranged from 0.33 to 1.60 with mean 0.77 (SD = 0.20), IV BIO5 from 0.57 to 2.26 with mean 1.01 (SD = 0.32), IV BIO10 from 0.63 to 1.76 with mean 0.98 (SD = 0.20), and IV BIO12 from 0.46 to 2.03 with mean 0.96 (SD = 0.28). The most variable IV was IV BIO5 (range 1.69) followed by IV BIO12 (range 1.57). Concerning soil variables, considerable variability among sites was found in volumetric percentage of coarse fragments (CRFVOL) and soil organic carbon content (ORCDRC), while other soil variables were more consistent among sites (Supplementary data Table S2).

4.6. Testing of Relationships Among Dormancy Traits, Geography and Environmental Variables

The matrix of environmental variables was checked for the presence of the multicollinearity using VIF. The reduced set of environmental variables (with VIF <15), including 14 bioclimatic variables and eight soil variables, was used in all further analyses. For each pair of variables, bivariate scatter plots together with fitted locally weighted smoothing were displayed and Pearson's Correlation Coefficient was calculated using the library Performance Analytics in R [90].

The matrix of the reduced set of environmental variables was analyzed by principal component analysis (PCA; [91]) using Canoco 5.10 [92] to find the main environmental gradients within the dataset. Several precipitation variables were log(x+1) transformed and subsequently each variable was standardized to zero mean and unit variance before PCA. A set of germination traits and geographic coordinates (latitude, longitude) were used as supplementary variables and correlated with the first two principal components. To control for possible spatial autocorrelation between each germination trait and principal components representing environmental gradients, a modified version of the *t*-test [93] was performed in SAM 4.0 (Rangel et al., 2010) [94]. To assess whether there is spatial autocorrelation present in the PCA scores along the first two axes and dormancy traits, Moran's I spatial correlation statistic [88] was calculated for each variable using PASSaGE v. 2.0 [95]. Ten distance classes with equal widths were created and Moran's I and its 95% CI were calculated for each distance class.

4.7. Phenotypic Plasticity by Macro-Environmental Clusters

We used accessions that had been tested over 3 years (Supplementary data Table S1) to calculate a norm of reaction for final PY dormancy (FPYD$_{25}$ and FPYD$_{35}$). First, selected *Medicago* accessions were grouped into four macro-environmental clusters (Supplementary data Table S3) based on Euclidean distance of environmental variables used for calculations of PCA. Agglomeration was performed using Ward's minimum-variance linkage algorithm. Before clustering, all variables were standardized to zero mean and unit variance. Second, a norm of reaction for each macro-environmental cluster was estimated as follows: each line represents the data for a different cluster and the effect of "environment" (treatments, 25/15 °C and 35/15 °C), separately for each experimental year [2]. To focus on the change of the trait in response to two temperature treatments we analyzed the FPYD means within each cluster per each year by ANOVA using the InfoStat software [96].

4.8. Genome-Wide Association Analysis

Genome-wide association analysis was performed on seven seed dormancy traits (FPYD$_{25}$, FPYD$_{35}$, AUC$_{25}$, AUC$_{35}$, AUC$_{35-25}$, PI$_{PY}$, PI$_{AUC}$) and three bioclimatic variables (BIO1, BIO9, BIO12) on 178 accessions. Prior to GWA analyses, normal distribution of each trait was checked using the Shapiro–Wilk test. Two contrasted algorithms were used to test markers–traits associations: Efficient Mixed Model Association (EMMA), a classical mixed linear model (MLM) for single locus GWAS [97], and FarmCPU, a multi-locus method combining the fixed effect model and random effect model iteratively in order to improve the statistical power of MLM methods [98]. Both algorithms were implemented in the R package rMVP using default parameters (P-value threshold 0.01) and run using a Single Nucleotide Polymorphism (SNP) dataset containing 5.85 million SNPs remapped in the *Medicago* genome v.5 [99]. Population structure, calculated using STRUCTURE by Bonhomme et al. [100], was used as a covariable. Normal distribution, QQ plots, and single/multiple Manhattan plots were performed using R package rMVP. *Medicago* genome version 5.0 of A17 genotype [101] was used to search for the encoded genes within the region of 10 kb from detected SNP. Transposable elements were excluded from the search. To link identified QTNs with putative causal gene by considering the linkage disequilibrium (LD), we selected all SNPs correlated ($r^2 > 0.7$) with the top identified QTNs within a 15kb genomic range, corresponding to the average LD block size present in the *Medicago* Hapmap population [100,102], and we listed gene IDs closely related to these SNPs. Seed expression pattern of the candidate genes was assessed using published *Medicago* seeds or seed coat expression studies [103,104] and web-based expression atlas [105]

5. Conclusions

We found that phenotypic plasticity of seed dormancy release was significantly correlated with increased gradient of aridity, suggesting that plastic responses to external stimuli provide seeds with bet-hedging capacity and the potential to cope with high levels of environmental heterogeneity. Genome-wide association analysis identified candidate genes associated with dormancy release. Gene ontology showed enrichment for genes involved in modification of the cell wall, as well as oxidative stress protection, mediating seed coat permeability and, ultimately, imbibition and germination. Knowledge of the seed dormancy regulation by environmental factors could be extended to other legume species, particularly to crop wild relatives of economically important species, such as chickpea, lentil, faba bean and soybean, as well as used in the management of endangered plant species with physical seed dormancy.

Supplementary Materials: The following are available online at http://www.mdpi.com/2223-7747/9/4/503/s1, Figure S1: Frequency distributions of dormancy traits, Figure S2: Correlation between the phenotypic plasticity index of final dormancy (PIPY) and selected environmental variables, Figure S3: Correlation chart of the dormancy release traits and environmental variables, Figure S4: Correlation chart of the dormancy release traits, soil variables and inter-annual climatic variables, Figure S5: Manhattan plots of mapped SNP markers associated with dormancy or bioclimatic traits, Figure S6: Quantile–quantile (Q-Q) plots for all the traits obtained by standard mixed linear model (EMMA) and multi-locus linear model (FarmCPU), Figure S7: Geographic distribution of studied M. truncatula accessions, Figure S8: Final PY dormancy (FPYD) of each year under two temperature treatments (35/15 °C and 25/15 °C). Different letters indicate significant differences among year for each temperature (Fisher's LSD test, $\alpha = 0.05$), Table S1: List of tested Medicago accessions with calculated seed dormancy traits and extracted environmental variables, Table S2: Basic descriptive statistics of 23 bioclimatic variables and 10 soil variables of sites of accessions origin, Table S3: Classification of 176 M. truncatula accessions in four cluster based on environmental and climatic conditions, Table S4: Pearson coefficients-probabilities between dormancy traits and bioclimatic variables, Table S5: Regression coefficient (r2) between environmental variables and plasticity index by macroecological and genetic clusters [106], Table S6: Complete list of QTN identified by GWA studies for each dormancy trait, Table S7: Over-representation analysis of the 136 candidate genes potentially involved in dormancy traits.

Author Contributions: Conceptualization, J.P.R. and P.S.; Data curation, J.B. and P.S.; Formal analysis, J.P.R., M.D., J.B., I.H., V.P., T.V., J.M., K.H., J.V. and P.S.; Funding acquisition, M.D. and P.S.; Investigation, J.P.R., J.B. and I.H.; Methodology, M.D., J.B., V.P., K.H. and P.S.; Project administration, I.H. and P.S.; Writing – original draft, J.P.R., M.D., J.B., J.V. and P.S.; Writing – review & editing, J.P.R., M.D., J.B., V.P., J.V. and P.S. All authors have read and agreed to the published version of the manuscript.

Funding: This research was funded by the Grant Agency of the Czech Republic [grant number 16-21053S].

Acknowledgments: We thank to Oldřich Trněný for construction of seed threshing device and Miroslav Hýbl for help with seeds cleaning.

Conflicts of Interest: The authors declare no conflict of interest. The funders had no role in the design of the study; in the collection, analyses, or interpretation of data; in the writing of the manuscript, or in the decision to publish the results.

Abbreviations

PY	physical dormancy
FPYD	final PY dormancy at 25/15 °C and 35/15 °C (FPYD$_{25}$, FPYD$_{35}$)
AUC	area under curve representing germination pattern (AUC$_{25}$, AUC$_{35}$)
PI$_{AUC}$	phenotypic plasticity index of the germination pattern
PI$_{PY}$	phenotypic plasticity index of final PY dormancy
IV	index of variability
FPYD$_M$	means of FPYD coefficients estimated for the two temperature treatments
AUC$_M$	means of AUC coefficients estimated for the two temperature treatments

References

1. Tognetti, P.M.; Mazia, N.; Ibáñez, G. Seed local adaptation and seedling plasticity account for *Gleditsia triacanthos* tree invasion across biomes. *Ann. Bot.* **2019**, *124*, 307–318. [CrossRef] [PubMed]
2. Valladares, F.; Sanchez-Gomez, D.; Zavala, M.A. Quantitative estimation of phenotypic plasticity: Bridging the gap between the evolutionary concept and its ecological applications. *J. Ecol.* **2006**, *94*, 1103–1116. [CrossRef]
3. Forsman, A. Rethinking phenotypic plasticity and its consequences for individuals, populations and species. *Heredity* **2015**, *115*, 276–284. [CrossRef] [PubMed]
4. Clausen, J. *Stages in the Evolution of Plant Species*; Cornell University Press: Ithaca, NY, USA, 1951.
5. Ungerer, M.; Johnson, L.; Herman, M. Ecological genomics: Understanding gene and genome function in the natural environment. *Heredity* **2008**, *100*, 178–183. [CrossRef]
6. Nicotra, A.B.; Atkin, O.K.; Bonser, S.P.; Davidson, A.M.; Finnegan, E.J.; Mathesius, U.; van Kleunen, M. Plant phenotypic plasticity in a changing climate. *Trends. Plant. Sci.* **2010**, *15*, 684–692. [CrossRef]
7. Richards, C.L.; Bossdorf, O.; Muth, N.Z.; Gurevitch, J.; Pigliucci, M. Jack of all trades, master of some? On the role of phenotypic plasticity in plant invasions. *Ecol. Lett.* **2006**, *9*, 981–993. [CrossRef]
8. Baskin, J.M.; Baskin, C.C. *Seeds: Ecology, Biogeography and Evolution of Dormancy and Germination*, 2nd ed.; Academic Press: Amsterdam, The Netherlands, 2014.
9. Willis, C.G.; Baskin, C.C.; Baskin, J.M.; Auld, J.R.; Venable, D.L.; Cavender-Bares, J.; Donohue, K.; Rubio de Casas, R. NESCent Germination Working Group. The evolution of seed dormancy: Environmental cues, evolutionary hubs, and diversification of the seed plants. *New. Phytol.* **2014**, *203*, 300–309. [CrossRef]
10. Hradilová, I.; Duchoslav, M.; Brus, J.; Pechanec, V.; Hýbl, M.; Kopecký, P.; Smržová, L.; Štefelová, N.; Vaclávek, T.; Bariotakis, M.; et al. Variation in wild pea (*Pisum sativum* subsp. *elatius*) seed dormancy and its relationship to the environment and seed coat traits. *PeerJ* **2019**, *7*, e6263.
11. Norman, H.C.; Cocks, P.C.; Galwey, N.W. Hardseededness in annual clovers: Variation between populations from wet and dry environments. *Aus. J. Agric. Sci.* **2002**, *53*, 821–829. [CrossRef]
12. Rosenberg, M.S.; Anderson, C.D. PASSaGE: Pattern Analysis, Spatial Statistics, and Geographic Exegesis. Version 2. *Method. Ecol. Evol.* **2011**, *2*, 229–232. [CrossRef]
13. Smýkal, P.; Vernoud, V.; Blair, M.W.; Soukup, A.; Thompson, R.D. The role of the testa during development and in establishment of dormancy of the legume seed. *Front. Plant. Sci.* **2014**, *5*, 351. [PubMed]
14. Sultan, S.E. Phenotypic plasticity for plant development, function and life history. *Trends. Plant. Sci.* **2000**, *5*, 537–542. [CrossRef]
15. Templeton, A.R.; Levin, D.A. Evolutionary consequences of seed pools. *Am. Nat.* **1979**, *114*, 232–249. [CrossRef]
16. Evans, M.E.K.; Ferrière, R.; Kane, M.J.; Venable, D.L. Bet hedging via seed banking in desert evening primroses (*Oenothera, Onagraceae*): Demographic evidence from natural populations. *Am. Nat.* **2007**, *169*, 184–194. [CrossRef] [PubMed]
17. Venable, D.L. Bet hedging in a guild of desert annuals. *Ecology* **2007**, *88*, 1086–1090. [CrossRef] [PubMed]
18. Duncan, C.; Schultz, N.L.; Good, M.K.; Lewandrowski, W.; Cook, S. The risk-takers and -avoiders: Germination sensitivity to water stress in an arid zone with unpredictable rainfall. *AOB Plants* **2019**. [CrossRef]
19. Clauss, M.J.; Venable, D.L. Seed germination in desert annuals: An empirical test of adaptive bet-hedging. *Am. Nat.* **2000**, *155*, 168–186. [CrossRef]

20. Van Klinken, R.D.; Flack, L.K.; Pettit, W. Wet-season Dormancy Release in Seed Banks of a Tropical Leguminous Shrub is Determined by Wet Heat. *Ann. Bot.* **2006**, *98*, 875–883. [CrossRef]

21. Batlla, D.; Benech-Arnold, R.L. A framework for the interpretation of temperature effects on dormancy and germination in seed populations showing dormancy. *Seed. Sci. Res.* **2015**, *25*, 147–158. [CrossRef]

22. Ellis, R.H.; Simon, G.; Covell, S. The influence of temperature on seed germination rate in grain legumes. III. A comparison of five faba bean genotypes at constant temperatures using a new screening method. *J. Exp. Bot.* **1987**, *38*, 1033–1043. [CrossRef]

23. Renzi, J.P.; Chantre, G.R.; Cantamutto, M.A. *Vicia villosa* ssp. villosa Roth field emergence model in a semiarid agroecosystem. *Grass. Forage. Sci.* **2018**, *73*, 146–158.

24. Rosbakh, S.; Poschlod, P. Initial temperature of seed germination as related to species occurrence along a temperature gradient. *Funct. Ecol.* **2015**, *29*, 5–14. [CrossRef]

25. Hudson, A.R.; Ayre, D.J.; Ooi, M.K.J. Physical dormancy in a changing climate. *Seed. Sci. Res.* **2015**, *25*, 66–81. [CrossRef]

26. Rubio de Casas, R.; Willis, C.G.; Pearse, W.D.; Baskin, C.C.; Baskin, J.M.; Cavender-Bares, J. Global biogeography of seed dormancy is determined by seasonality and seed size: A case study in the legumes. *New. Phytol.* **2017**, *214*, 1527–1536. [CrossRef]

27. Quinlivan, B.J. The effect of constant and fluctuating temperatures on the permeability of the hard seeds of some legume species. *Aust. J. Agric. Res.* **1961**, *12*, 1009–1022. [CrossRef]

28. Quinlivan, B.J. The relationship between temperature fluctuations and the softening of hard seeds of some legume species. *Aust. J. Agric. Res.* **1966**, *17*, 625–631. [CrossRef]

29. Quinlivan, B.J.; Millington, A.J. The effect of a Mediterranean summer environment on the permeability of hard seeds of subterranean clover. *Aust. J. Agric. Res.* **1962**, *13*, 377–387. [CrossRef]

30. Berger, J.D.; Shrestha, D.; Ludwig, C. Reproductive Strategies in Mediterranean Legumes: Trade-Offs between Phenology, Seed Size and Vigor within and between Wild and Domesticated Lupinus Species Collected along Aridity Gradients. *Front. Plant. Sci.* **2017**, *8*, 548. [CrossRef]

31. Ronfort, J.; Bataillon, T.; Santoni, S.; Delalande, M.; David, J.L.; Prosperi, J.-M. Microsatellite diversity and broad scale geographic structure in a model legume: Building a set of nested core collection for studying naturally occurring variation in *Medicago truncatula*. *BMC Plant Biol.* **2006**, *6*, 28. [CrossRef]

32. Gallardo, K.; Firnhaber, C.; Zuber, H.; Héricher, D.; Belghazi, M.; Henry, C.; Küster, H.; Thompson, R. A combined proteome and transcriptome analysis of developing *Medicago truncatula* seeds: Evidence for metabolic specialization of maternal and filial tissues. *Mol. Cell. Proteom.* **2007**, *6*, 2165–2179. [CrossRef]

33. Bolingue, W.; Vu, B.L.; Leprince, O.; Buitink, J. Characterization of dormancy behaviour in seeds of the model legume *Medicago truncatula*. *Seed. Sci. Res.* **2010**, *20*, 97–107. [CrossRef]

34. Faria, J.M.R.; Buitink, J.; van Lammeren, A.A.M.; Hilhors, H.W.M. Changes in DNA and microtubules during loss and re-establishment of desiccation tolerance in germinating *Medicago truncatula* seeds. *J. Exp. Bot.* **2005**, *56*, 2119–2130. [CrossRef] [PubMed]

35. Vu, W.T.; Chang, P.L.; Moriuchi, K.S.; Friesen, M.L. Genetic variation of transgenerational plasticity of offspring germination in response to salinity stress and the seed transcriptome of *Medicago truncatula*. *BMC Evol. Biol.* **2015**, *15*, 59. [CrossRef] [PubMed]

36. Stanton-Geddes, J.; Paape, T.; Epstein, B.; Briskine, R.; Yoder, J.; Mudge, J.; Bharti, A.K.; Farmer, A.D.; Zhou, P.; Denny, R.; et al. Candidate genes and genetic architecture of symbiotic and agronomic traits revealed by whole-genome, sequence-based association genetics in *Medicago truncatula*. *PLoS ONE* **2013**, *8*, e65688. [CrossRef] [PubMed]

37. Burgarella, C.; Chantret, N.; Gay, L.; Prosperi, J.-M.; Bonhomme, M.; Tiffin, P.; Young, N.D.; Ronfort, J. Adaptation to climate through flowering phenology: A case study in *Medicago truncatula*. *Mol. Ecol.* **2016**, *25*, 3397–3415. [CrossRef]

38. Julier, B.; Huguet, T.; Chardon, F.; Ayadi, R.; Pierre, J.-B.; Prosperi, J.-M.; Barre, P.; Huyghe, C. Identification of quantitative trait loci influencing aerial morphogenesis in the model legume *Medicago truncatula*. *Appl. Genet.* **2007**, *114*, 1391–1406. [CrossRef]

39. Thompson, J.D. *Plant Evolution in the Mediterranean*; Oxford University Press: Oxford, UK, 2005.

40. Finch-Savage, W.E.; Footitt, S. Seed dormancy cycling and the regulation of dormancy mechanisms to time germination in variable field environments. *J. Exp. Bot.* **2017**, *68*, 843–856. [CrossRef]

41. Tribouillois, H.; Dürr, C.; Demilly, D.; Wagner, M.-H.; Justes, E. Determination of Germination Response to Temperature and Water Potential for a Wide Range of Cover Crop Species and Related Functional Groups. *PLoS ONE* **2016**, *11*, e0161185. [CrossRef]

42. Hu, X.W.; Fan, Y.; Baskin, C.C.; Baskin, J.M.; Wang, Y.R. Comparison of the effects of temperature and water potential on seed germination of Fabaceae species from desert and subalpine grassland. *Am. J. Bot.* **2015**, *102*, 649–660. [CrossRef]

43. Long, R.L.; Gorecki, M.J.; Renton, M.; Scott, J.K.; Colville, L.; Goggin, D.E.; Commander, L.E.; Westcott, D.A.; Cherry, H.; Finch-Savage, W.E. The ecophysiology of seed persistence: A mechanistic view of the journey to germination or demise. *Biol. Rev. Camb. Philos. Soc.* **2015**, *90*, 31–59. [CrossRef]

44. Ferreras, A.E.; Marcora, P.I.; Venier, M.P.; Funes, G. Different strategies for breaking physical seed dormancy in field conditions in two fruit morphs of *Vachellia caven* (*Fabaceae*). *Seed. Sci. Res.* **2018**, *28*, 8–15. [CrossRef]

45. Debieu, M.; Tang, C.; Stich, B.; Sikosek, T.; Effgen, S.; Josephs, E.; Schmitt, J.; Nordborg, M.; Koornneef, M.; Meaux, J. Co-Variation between Seed Dormancy, Growth Rate and Flowering Time Changes with Latitude in *Arabidopsis thaliana*. *PLoS ONE* **2013**, *8*, e61075. [CrossRef] [PubMed]

46. Postma, F.M.; Ågren, J. Maternal environment affects the genetic basis of seed dormancy in *Arabidopsis thaliana*. *Mol. Ecol.* **2015**, *24*, 785–797. [CrossRef] [PubMed]

47. Lampei, C.; Metz, J.; Tielbörger, K. Clinal population divergence in an adaptive parental environmental effect that adjusts seed banking. *New. Phytol.* **2017**, *214*, 1230–1244. [CrossRef]

48. Wagmann, K.; Hautekèete, N.-C.; Piquot, Y.; Meunier, C.; Schmitt, S.E.; Van Dijk, H. Seed dormancy distribution: Explanatory ecological factors. *Ann. Bot.* **2012**, *110*, 1205–1219. [CrossRef]

49. Burghardt, L.T.; Metcalf, C.J.E.; Donohue, K. A cline in seed dormancy helps conserve the environment experienced during reproduction across the range of *Arabidopsis thaliana*. *Am. J. Bot.* **2016**, *103*, 47–59. [CrossRef]

50. Cochrane, A. Multi-year sampling provides insight into the bet-hedging capacity of the soil-stored seed reserve of a threatened Acacia species from Western Australia. *Plant. Ecol.* **2019**, *220*, 241–253. [CrossRef]

51. Ten Brink, H.; Gremer, J.R.; Kokko, H. Optimal germination timing in unpredictable environments: The importance of dormancy for both among- and within-season variation. *Ecol. Lett.* **2020**, *23*, 620–630. [CrossRef]

52. Taylor, G.B. Hardseededness in Mediterranean annual pasture legumes in Australia: A review. *Aust. J. Agric. Res.* **2005**, *56*, 645–661. [CrossRef]

53. Jaganathan, G.K.; Dalrymple, S.E.; Liu, B. Towards an Understanding of Factors Controlling Seed Bank Composition and Longevity in the Alpine Environment. *Bot. Rev.* **2015**, *81*, 70–103. [CrossRef]

54. Tozer, M.G.; Ooi, M.K.J. Humidity-regulated dormancy onset in the Fabaceae: A conceptual model and its ecological implications for the Australian wattle *Acacia saligna*. *Ann. Bot.* **2014**, *114*, 579–590. [CrossRef] [PubMed]

55. Penfield, S.; MacGregor, D.R. Effects of environmental variation during seed production on seed dormancy and germination. *J. Exp. Bot.* **2017**, *68*, 819–825. [CrossRef] [PubMed]

56. Stine, P.A.; Hunsaker, C.T. An introduction to uncertainty issues for spatial data used in ecological applications. In *Spatial Uncertainty in Ecology*; Springer: New York, NY, USA, 2001; pp. 91–107.

57. Brus, J.; Pechanec, V.; Machar, I. Depiction of uncertainty in the visually interpreted land cover data. *Ecol. Inf.* **2018**, *17*, 10–13. [CrossRef]

58. Buisson, L.; Thuiller, W.; Casajus, N.; Lek, S.; Grenouillet, G. Uncertainty in ensemble forecasting of species distribution. *Glob. Chang. Biol.* **2010**, *16*, 1145–1157. [CrossRef]

59. Meyer, C.; Weigelt, P.; Kreft, H. Multidimensional biases, gaps and uncertainties in global plant occurrence information. *Ecol. Lett.* **2016**, *19*, 992–1006. [CrossRef] [PubMed]

60. Kerdaffrec, E.; Nordborg, M. The maternal environment interacts with genetic variation in regulating seed dormancy in Swedish *Arabidopsis thaliana*. *PLoS ONE* **2017**, *12*, e0190242. [CrossRef] [PubMed]

61. Dias, P.M.B.; Brunel-Muguet, S.; Dürr, C.; Huguet, T.; Demilly, D.; Wagner, M.-H.; Teulat-Merah, B. QTL analysis of seed germination and pre-emergence growth at extreme temperatures in *Medicago truncatula*. *Appl. Genet.* **2011**, *122*, 429–444. [CrossRef]

62. Kang, Y.; Sakiroglu, M.; Krom, N.; Stanton-Geddes, J.; Wang, M.; Lee, Y.-C.; Young, N.D.; Udvardi, M. Genome-wide association of drought-related and biomass traits with HapMap SNPs in *Medicago truncatula*. *Plant Cell Environ.* **2015**, *38*, 1997–2011. [CrossRef]

63.   Francoz, E.; Lepiniec, L.; North, H.M. Seed coats as an alternative molecular factory: Thinking outside the box. *Plant. Reprod.* **2018**, *31*, 327–342. [CrossRef]

64.   Hradilová, I.; Trněný, O.; Válková, M.; Cechová, M.; Janská, A.; Prokešová, L.; Aamir, K.; Krezdorn, N.; Rotter, B.; Winter, P.; Varshney, R.K.; et al. A combined comparative transcriptomic, metabolomic, and anatomical analyses of two key domestication traits: Pod dehiscence and seed dormancy in pea (*Pisum. sp.*). *Front. Plant. Sci.* **2017**, *8*, 542e.

65.   Leubner-Metzger, G. Functions and regulation of β-1,3-glucanases during seed germination, dormancy release and after-ripening. *Seed Sci. Res.* **2003**, *13*, 17–34. [CrossRef]

66.   Leubner-Metzger, G.; Frundt, C.; Vogeli-Lange, R.; Meins, F. Class I [beta]-1,3-Glucanases in the Endosperm of Tobacco during Germination. *Plant. Physiol.* **1995**, *109*, 751–759. [CrossRef] [PubMed]

67.   Nighojkar, A.; Srivastava, S.; Kumar, A. Pectin methylesterase from germinating *Vigna. sinensis.* seeds. *Plant. Sci.* **1994**, *103*, 115–120. [CrossRef]

68.   Sitrit, Y.; Hadfield, K.A.; Bennett, A.B.; Bradford, K.J.; Downie, A.B. Expression of a Polygalacturonase Associated with Tomato Seed Germination. *Plant. Physiol.* **1999**, *121*, 419–428. [CrossRef]

69.   Buckeridge, M.S.; Hutcheon, I.S.; Reid, J.S.G. The Role of Exo-(1→4)-β-galactanase in the Mobilization of Polysaccharides from the Cotyledon Cell Walls of *Lupinus. angustifolius.* Following Germination. *Ann. Bot.* **2005**, *96*, 435–444. [CrossRef]

70.   Hancock, J.T.; Neill, S.J. Nitric Oxide: Its Generation and Interactions with Other Reactive Signaling Compounds. *Plants* **2019**, *8*, 41. [CrossRef]

71.   Müller, K.; Linkies, A.; Vreeburg, R.A.M.; Fry, S.C.; Krieger-Liszkay, A.; Leubner-Metzger, G. In vivo cell wall loosening by hydroxyl radicals during cress seed germination and elongation growth. *Plant. Physiol.* **2009**, *150*, 1855–1865. [CrossRef]

72.   Jeevan Kumar, S.P.; Rajendra Prasad, S.; Banerjee, R.; Thammineni, C. Seed birth to death: Dual functions of reactive oxygen species in seed physiology. *Ann. Bot.* **2015**, *116*, 663–668. [CrossRef]

73.   Raviv, B.; Godwin, J.; Granot, G.; Grafi, G. The Dead Can Nurture: Novel Insights into the Function of Dead Organs Enclosing Embryos. *Int. J. Mol. Sci.* **2018**, *19*, 2455. [CrossRef]

74.   Bewley, J.D.; Bradford, K.J.; Hilhorst, H.W.M.; Nonogaki, H. *Seeds: Physiology of Development, Germination and Dormancy*; Springer: New York, NY, USA, 2013.

75.   Novo-Uzal, E.; Fernández-Pérez, F.; Herrero, J.; Gutiérrez, J.; Gómez-Ros, L.V.; Bernal, M.Á.; Díaz, J.; Cuello, J.; Pomar, F.; Pedreño, M.Á. From Zinnia to *Arabidopsis*: Approaching the involvement of peroxidases in lignification. *J. Exp. Bot.* **2013**, *64*, 3499–3518. [CrossRef]

76.   Pollard, A.T. Seeds vs fungi: An enzymatic battle in the soil seedbank. *Seed Sci. Res.* **2018**, *28*, 197–214. [CrossRef]

77.   Graeber, K.; Nakabayashi, K.; Miatton, E.; Leubner-Metzger, G.; Soppe, W.J. Molecular mechanisms of seed dormancy. *Plant. Cell. Environ.* **2012**, *35*, 1769–86. [CrossRef] [PubMed]

78.   Yoder, J.B.; Stanton-Geddes, J.; Zhou, P.; Briskine, R.; Young, N.D.; Tiffin, P. Genomic signature of adaptation to climate in *Medicago truncatula*. *Genetics* **2014**, *196*, 1263–1275. [CrossRef] [PubMed]

79.   De Boor, C. *A Practical Guide to Splines*; Springer: New York, NY, USA, 1978.

80.   Machalová, J. Optimal interpolating and optimal smoothing spline. *J. Elect. Eng.* **2002**, *5312*, 79–82.

81.   Kader, M.A. A Comparison of Seed Germination Calculation Formulae and the Associated Interpretation of Resulting Data. *J. Proc. R. Soc. New. South. Wales.* **2005**, *138*, 65–75.

82.   Talská, R.; Machalová, J.; Smýkal, P.; Hron, K. A comparison of seed germination coefficients using functional regression. *Appl. Plant. Sci.* **2020**, (in press).

83.   ESRI. Available online: https://pro.arcgis.com (accessed on 5 October 2019).

84.   WorldClim. Available online: http://worldclim.org (accessed on 5 October 2019).

85.   Naimi, B. Uncertainty Analysis for Species Distribution Models. Version 1.1-18. 2017. Available online: http://r-gis.net (accessed on 5 October 2019).

86.   Fick, S.E.; Hijmans, R.J. WorldClim 2: New 1-km spatial resolution climate surfaces for global land areas. *Int. J. Climatol.* **2017**, *37*, 4302–4315. [CrossRef]

87.   Cobon, D.H.; Kouadio, L.; Mushtaq, S.; Jarvis, C.; Carter, J.; Stone, G.; Davis, P. Evaluating the shifts in rainfall and pasture-growth variabilities across the pastoral zone of Australia during 1910–2010. *Crop. Pasture. Sci.* **2019**, *70*, 634–647. [CrossRef]

88. Vega, G.C.; Pertierra, L.R.; Olalla-Tarraga, M.A. MERRAclim, a high-resolution global dataset of remotely sensed bioclimatic variables for ecological modelling. *Sci. Data* **2018**, *5*, 170078. [CrossRef]

89. ISRIC. Available online: https://www.isric.org/explore/soilgrids (accessed on 5 October 2019).

90. Hengl, T.; de Mendes, J.J.; Heuvelink, G.B.M.; Ruiperez Gonzalez, M.; Kilibarda, M.; Blagotić, A.; Shangguan, W.; Wright, M.N.; Geng, X.; Bauer-Marschallinger, B.; et al. SoilGrids250m: Global gridded soil information based on machine learning. *PLoS ONE* **2017**, *12*, e0169748. [CrossRef]

91. Peterson, B.G. Econometric Tools for Performance and Risk Analysis. 2019. Available online: https: //github.com/braverock/PerformanceAnalytics (accessed on 10 October 2019).

92. Legendre, P.; Legendre, L. *Numerical Ecology*; Elsevier: Amsterdam, The Netherlands, 2012; p. 990.

93. Ter Braak, C.J.F.; Šmilauer, P. *Canoco Reference Manual and User's Guide: Software for Ordination*, version 5.0; Microcomputer Power: Ithaca, NY, USA, 2012; p. 496.

94. Dutilleul, P.; Clifford, P.; Richardson, S.; Hemon, D. Modifying the t test for assessing the correlation between two spatial processes. *Biometrics* **1993**, *49*, 305–314. [CrossRef]

95. Rangel, T.F.; Diniz-Filho, J.A.F.; Bini, L.M. SAM: A comprehensive application for Spatial Analysis in Macroecology. *Ecography* **2010**, *33*, 46–50. [CrossRef]

96. Di Rienzo, J.A.; Casanoves, F.; Balzarini, M.G.; Gonzalez, L.; Tablada, M.; Robledo, C.W. *InfoStat Version 2013*; Grupo InfoStat, FCA; Universidad Nacional de Córdoba: Córdoba, Argentina, 2013.

97. Yu, J.; Buckler, E.S. Genetic association mapping and genome organization of maize. *Curr. Opin. Biotech.* **2006**, *17*, 155–160. [CrossRef] [PubMed]

98. Liu, X.; Huang, M.; Fan, B.; Buckler, E.S.; Zhang, Z. Iterative Usage of Fixed and Random Effect Models for Powerful and Efficient Genome-Wide Association Studies. *PLoS Genet.* **2016**, *16*, e1005767. [CrossRef] [PubMed]

99. Pecrix, Y.; Staton, S.E.; Sallet, E.; Lelandais-Brière, C.; Moreau, S.; Carrère, S.; Zahm, M.; Kreplak, J.; Mayjonade, B.; Carine Satgé, C.; et al. Whole-genome landscape of *Medicago. truncatula.* symbiotic genes. *Nat. Plants.* **2018**, *4*, 1017–1025. [CrossRef]

100. Bonhomme, M.; André, O.; Badis, Y.; Ronfort, J.; Burgarella, C.; Chantret, N.; Miteul, H.; Hajri, A.; Baranger, A.; Tiffin, P.; et al. High-density genome-wide association mapping implicates an F-box encoding gene in Medicago truncatula resistance to Aphanomyces euteiches. *New. Phytol.* **2014**, *201*, 1328–1342.

101. Medicago Truncatula A17 r5.0 Genome Portal. 2019. Available online: https://medicago.toulouse.inra.fr/ MtrunA17r5.0-ANR (accessed on 15 December 2019).

102. Branca, A.; Paape, T.D.; Zhou, P.; Briskine, R.; Farmer, A.D.; Mudge, J.; Ben, C.; Denny, R.; Sadowsky, M.J.; Ronfort, J.; et al. Whole-genome nucleotide diversity, recombination, and linkage disequilibrium in the model legume Medicago truncatula. *Proc. Nat. Acad. Sci. USA* **2011**, *108*, E864–E870.

103. Fu, F.; Zhang, W.; Li, Y.Y.; Wang, H.L. Establishment of the model system between phytochemicals and gene expression profiles in Macroscleid cells of *Medicago truncatula*. *Sci. Rep.* **2017**, *7*, 2580. [CrossRef]

104. Verdier, J.; Dessaint, F.; Schneider, C.; Abirached-Darmency, M. A combined histology and transcriptome analysis unravels novel questions on *Medicago truncatula* seed coat. *J. Exp. Bot.* **2013**, *64*, 459–470. [CrossRef]

105. Bar. University of Toronto. Available online: http://bar.utoronto.ca/efpmedicago/cgi-bin/efpWeb.cgi (accessed on 15 December 2019).

106. Gentzbittel, L.; Ben, C.; Mazurier, M.; Shin, M.-G.; Lorenz, T.; Rickauer, M.; Marjoram, P.; Nuzhdin, S.V.; Tatarinova, T.V. WhoGEM: An admixture-based prediction machine accurately predicts quantitative functional traits in plants. *Genome. Biol.* **2019**, *20*, 106. [CrossRef]

Article

Effects of GABA and Vigabatrin on the Germination of Chinese Chestnut Recalcitrant Seeds and Its Implications for Seed Dormancy and Storage

Changjian Du [1,†], Wei Chen [1,†], Yanyan Wu [1], Guangpeng Wang [2], Jiabing Zhao [3], Jiacheng Sun [1], Jing Ji [1], Donghui Yan [4], Zeping Jiang [4] and Shengqing Shi [1,*]

1   State Key Laboratory of Tree Genetics and Breeding, Key Laboratory of Tree Breeding and Cultivation of State Forestry and Grassland Administration, Research Institute of Forestry, the Chinese Academy of Forestry, 1958 Box, Beijing 100091, China; duchnagjian@126.com (C.D.); 15764356257@163.com (W.C.); 15732153437@163.com (Y.W.); sjc9867@163.com (J.S.); jijing0401@163.com (J.J.)

2   Institute for Pomology, Hebei Academy of Agriculture and Forestry Sciences, Changli 066600, China; wangguangpeng430@163.com

3   College of Forestry, Hebei Agricultural University, Baoding 071001, China; zhaojiabing0310@163.com

4   Key Laboratory of Forest Ecology and Environment of National Forestry and Grassland Administration, Research Institute of Forest Ecology, Environment and Protection, the Chinese Academy of Forestry, 1958 Box, Beijing 100091, China; yandh@caf.ac.cn (D.Y.); jiangzp@caf.ac.cn (Z.J.)

*   Correspondence: shi.shengqing@caf.ac.cn; Tel.: +86-10-62889054
†   These authors contributed equally.

Received: 19 March 2020; Accepted: 1 April 2020; Published: 3 April 2020

Abstract: Recalcitrant chestnut seeds are rich in γ-aminobutyric acid (GABA), which negatively regulates adventitious root development by altering carbon/nitrogen metabolism. However, little is known regarding the role of this metabolite in chestnut seeds. In this study, we investigated the effects of GABA changes on the germination of chestnut seeds treated with exogenous GABA and vigabatrin (VGB, which inhibits GABA degradation). Both treatments significantly inhibited seed germination and primary root growth and resulted in the considerable accumulation of H_2O_2, but the endogenous GABA content decreased before germination at 48 h. Soluble sugar levels increased before germination, but subsequently decreased, whereas starch contents were relatively unchanged. Changes to organic acids were observed at 120 h after sowing, including a decrease and increase in citrate and malate levels, respectively. Similarly, soluble protein contents increased at 120 h, but the abundance of most free amino acids decreased at 48 h. Moreover, the total amino acid levels increased only in response to VGB at 0 h. Accordingly, GABA and VGB altered the balance of carbon and nitrogen metabolism, thereby inhibiting chestnut seed germination. These results suggested that changes to GABA levels in chestnut seeds might prevent seed germination. The study data may also help clarify the dormancy and storage of chestnut seeds, as well as other recalcitrant seeds.

Keywords: chestnut; GABA; seed germination; carbon metabolism; nitrogen metabolism

1. Introduction

Chestnut (genus *Castanea*; family Fagaceae), which is a major nut crop in East Asia and Southern Europe, is unique among temperate nut crops because its seed is starchy rather than oily, making it ecologically and economically valuable [1,2]. The production of American chestnut, however, has been severely affected in North America by chestnut blight due to *Cryphonectria parasitica* [3,4]. As a perennial crop, chestnut is not only used as an important starch-based food product consumed by people living in rural areas [5,6], but also as a potential functional food because it is a rich source of bioactive compounds, including phenolics [5]. Additionally, chestnut seeds contain a considerable abundance of γ-aminobutyric acid (GABA) [7], which is the key component of the GABA shunt crucial for carbon and nitrogen metabolism in plants [8]. Previous studies have also identified GABA-enriched functional foods [9,10]. However, there is still relatively little information regarding GABA functions affecting chestnut seed activities, including germination.

Seeds, including orthodox and recalcitrant seeds [11], play a major role in agriculture, serving as food and feed, as well as plant propagation units. Seed germination is influenced by internal metabolic changes [12,13] and a complex process in which starch-degrading α-amylase and various proteases are activated to decrease the total dry matter content [14]. During germination, starch and proteins are degraded into smaller molecules, such as soluble sugars and free amino acids [15], resulting in a significant increase in the free amino acid content [16]. Previous studies proved that GABA levels increase during the germination of barnyard millet [16] and wheat [17,18] seeds under normal conditions, as well as wheat seeds under saline conditions [18]. The accumulation of endogenous GABA in dry seeds facilitates early metabolic reorganization during germination [19]. Moreover, exogenous GABA reportedly affects the seed germination process in barley [20] and *Haloxylon ammodendron* [21] under normal conditions. It also modulates respiration during the germination of *H. ammodendron* seeds [21] and regulates H_2O_2 production in *Caragana intermedia* [22] and poplar [23] in response to salt stress. The application of exogenous GABA can also mitigate salt-mediated damages by enhancing starch catabolism and the use of sugars and amino acids [24] or by enhancing the antioxidant system to induce the accumulation of phenolic compounds during seed germination [25]. Additionally, an exogenous GABA treatment can delay the loss of titratable acidity and malate, thereby maintaining the quality of stored apple fruits [26]. These studies confirmed that GABA can affect seed germination or fruit storage by altering the metabolism of carbon and nitrogen, as well as reactive oxygen species (ROS).

Chestnut seeds are recalcitrant and exhibit dormancy [27,28], which differentiates them from most other recalcitrant seeds [29]. This dormancy increases the shelf-life of the seeds used as food and enables the seeds to survive unfavorable winter conditions so they can germinate in the following spring. The normal germination of chestnuts is vital for their use as rootstock seedlings during grafting [30]. Earlier reports indicated that the germination rates of 56 selected types and/or cultivars of European chestnut seeds were 17.6–86.6% [31] and 51.6–97.3% [32] in two regions in Turkey, depending on their dormancy-breaking time due to stratification temperatures [28,31,32]. Similar to orthodox seeds, the dormancy of recalcitrant seeds appears to be induced by abscisic acid (ABA) [29,33], which is abundant in chestnut seed coats [34]. In contrast, the contents of the ABA antagonist GA_3 are high in embryos and cotyledons [34]. Thus, different exogenous treatments have been used to increase germination rates, including an H_2O_2 treatment [35] and the application of GA_3 [30,34]. Notably, an earlier investigation demonstrated that GABA is a major amino component associated with the high accumulation of several amino acids during chestnut seed germination [36], suggesting there is a close relationship between GABA and the recalcitrant chestnut seeds.

The effects of GABA on chestnut seed germination remain unclear. Thus, on the basis of our previous findings that GABA regulates stress responses [22,23] and adventitious root development [37], in this study, we treated Chinese chestnut (*Castanea mollissima*) cultivar "Yanshanzaofeng" seeds with GABA and vigabatrin (VGB; a specific GABA transaminase inhibitor) to investigate seed germination changes, as well as the central carbon/nitrogen metabolic activities, which demonstrated that both treatments inhibited chestnut seed germination and might by altering the balance of carbon and nitrogen metabolism, which would provide a better understanding for elucidating the role of GABA during the storage and germination of recalcitrant chestnut seeds.

2. Materials and Methods

2.1. Plant Materials and Treatments

Chinese chestnut (*Castanea mollissima*) cultivar "Yanshanzaofeng" seeds were collected from healthy trees growing in the core chestnut-producing region of Qianxi county, Tangshan city, Hebei province, China, after which they were stored at 0–1 °C until further use. Relatively uniform chestnut seeds were washed five times with sterile water and air-dried. The seeds were then soaked in sterile water (control/CK), 10 mM GABA (lab use only; Sigma-Aldrich, St. Louis, MO, USA), or 100 µM vigabatrin (VGB; lab use only; MedChem Express, Monmouth Junction, NJ, USA) for 15 h at 25 °C before they were placed evenly on a tray (33.5 cm × 26 cm × 11 cm) containing sterilized sand. A germination test was conducted in a climate chamber with a 16 h light (25 °C): 8 h dark (20 °C) cycle and 60% relative humidity. The seed germination rate was calculated, and the root length was measured after 2, 5, 8, 15, and 30 days. Five seeds per replicate were collected at 15 h before the treatment (t0; i.e., the time-point when the seed imbibition was initiated) and at 0, 48, and 120 h after sowing. After discarding the seed coats, the kernels were ground into a powder and stored at −80 °C for the subsequent physiological measurements. Each treatment was completed with three replicates, each comprising 50 seeds.

2.2. Calculation of the Seed Germination Rate

Seed radicle protrusion was used as the criterion for judging germination. Germinated seeds were counted at the designated treatment times, after which the germination rate was calculated based on the ratio of the number of germinated seeds and the total number of seeds in each treatment.

2.3. Measurement of Reactive Oxygen Species

The H_2O_2 content was measured as previously described [38]. Briefly, 0.1 g fresh powder were dissolved in 1 mL acetone and then thoroughly mixed on ice. After a centrifugation ($8000\times g$ for 10 min at 4 °C), the supernatant was mixed with a titanium sulfate solution and concentrated ammonia. After another centrifugation ($4000\times g$ for 10 min at 25 °C), the sediment was dissolved in concentrated sulfuric acid and incubated at room temperature for 5 min. The absorbance of the reaction solution was measured at 415 nm, and the H_2O_2 content was recorded as µmol/g fresh weight (FW).

2.4. Measurement of Soluble Sugars and Starch

The total soluble sugar content was measured with a commercial assay kit (Comin Biotechnology, Suzhou, China) based on the anthrone-sulfuric acid method as previously described [39]. Briefly, 0.1 g fresh powder were dissolved in 1 mL sterile water and heated for 10 min at 95 °C. After a centrifugation ($8000\times g$ for 10 min at 25 °C), the diluted supernatant was added to a reaction mixture comprising anthranone in ethyl acetate and a concentrated sulfuric acid solution and then heated for 10 min at 95 °C. The absorbance of the reaction solution was measured at 620 nm, and the soluble sugar content was recorded as % FW.

The starch content was quantified with a commercial assay kit (Comin Biotechnology) based on a previously described method [40]. Briefly, 0.1 g fresh powder were dissolved in 1 mL ethanol solution and heated at 80 °C for 30 min. The solution was centrifuged ($3000\times g$ for 5 min at 25 °C), and the sediment was dissolved in 0.5 mL sterile water and heated at 95 °C for 15 min. After cooling, 0.35 mL perchloric acid and 0.85 mL sterile water were added, and the resulting solution was thoroughly mixed and then centrifuged ($3000\times g$ for 10 min at 25 °C). A 50 µL aliquot of the solution was mixed with 250 µL reaction mixture (3.75 mL anthranone solution and 21.25 mL concentrated sulfuric acid) and then heated at 80 °C for 10 min. The absorbance of the reaction solution was measured at 620 nm, and the starch content was recorded as % FW.

2.5. Measurement of Soluble Proteins and Total Amino Acids

The total soluble protein content was measured with a Bradford Protein Assay Kit (Sangon Biotech, Shanghai, China) based on the Coomassie brilliant blue method. Briefly, 0.1 g fresh powder were suspended in 5 mL phosphate buffer (pH 7.2) and centrifuged ($8000\times g$ for 10 min at 4 °C). The diluted supernatant was mixed with Coomassie brilliant blue R-250, after which the absorbance was measured at 595 nm. The total soluble protein content was recorded as mg/g FW.

The total amino acid content was measured with a commercial assay kit (Comin Biotechnology) based on a previously described method [41]. Briefly, 0.1 g fresh powder were dissolved in 1 mL glacial acetic acid, mixed thoroughly, and boiled for 15 min. After a centrifugation ($8000 \times g$ for 5 min at 4 °C), the supernatant was added to a reaction solution consisting of 100 μL sodium acetate-glacial acetic acid, 100 μL ninhydrin, and 10 μL ascorbic acid. After boiling for 15 min, the absorbance of the reaction solution was measured at 570 nm, and the total amino acid content was recorded as mg/g FW.

2.6. Measurement of Organic Acids and Amino Acids

To analyze the organic acid and amino acid contents, 0.4 g fresh powder were suspended in 2 mL sterile water and boiled at 100 °C for 90 min. After a centrifugation, the supernatant was filtered (0.45 μm pore size). The organic acid and amino acid contents of the filtrate were analyzed by high-performance liquid chromatography with the 2695 Separations Module (Waters, Milford, MA, USA) as previously described [37,42]. All assays were completed with three biological replicates per treatment.

Organic acids: The citrate, succinate, and malate contents of the prepared extract were measured at 214 nm with the Waters 2487 UV detector, a Thermo ODS HYPERSIL (4.6 mm × 200 mm) chromatographic column, and 3% CH_3OH + 97% H_2O as the mobile phase (flow rate, 0.8 mL/min). The organic acid contents were recorded as mg/g FW.

Amino acids: The prepared extract was analyzed with a pre-column derivatized with the AccQ Fluor reagent. The mobile phase consisted of 140 mM NaAc solution (containing 17 mM triethylamine, pH 4.95) and a 60% aqueous acetonitrile solution. A Waters 2475 fluorescence detector and an AccQ Tag amino acid analysis column (3.9 mm × 150 mm) were used. The separate amino acids were detected at excitation and emission wavelengths of 250 and 395 nm, respectively. The amino acid contents were recorded as μg/g FW.

GABA: The prepared extract was mixed with acetonitrile containing 1% 2,4-dinitrofluorobenzene, after which an equal volume of $NaHCO_3$ (pH 9) buffer was added, and the solution was incubated for 1 h at 60 °C. The solution contents were separated on a C_{18} SunFire column (4.6 mm × 250 mm), and the eluted products were detected with the Waters 2487 UV detector. The column temperature was maintained at 35 °C, and the mobile phase comprised phosphate buffer (pH 7), water, and acetonitrile, with a flow rate of 0.1 mL/min. The separated GABA was detected at 360 nm. The GABA content was recorded as μg/g FW.

2.7. Statistical Analysis

Data were compared and analyzed with ANOVA (analysis of variance), and multiple comparisons were made with SPSS 16.0 (SPSS, Chicago, IL, USA). Differences were scored as significant at the $p < 0.05$ or $p < 0.01$ levels. A principal component analysis (PCA) was performed with the command prcomp() in R (http://www.r-project.org/), as previously described [43,44].

3. Results

3.1. Effects of GABA and VGB on Seed Germination Characteristics

The CK, 10 mM GABA, and 100 μM VGB solutions did not induce seed germination on Day 2. The germination rate of the CK seeds increased after five days, whereas the 10 mM GABA and 100 μM VGB treatments inhibited chestnut seed germination and primary root growth at five days after sowing (Figure 1A). The germination rates of the CK seeds gradually increased by over 27.8–54.8% between Days 5 and 30 (Figure 1B), whereas the germination rates gradually decreased by over 11.0–17.2% and 17.7–21.1% following the GABA and VGB treatments, respectively (Figure 1B). The VGB treatment decreased the average root length per seed between Days 8 and 30 (Figure 1C), whereas the total root length was generally decreased by both GABA and VGB between Days 5 and 30 ($p < 0.05$). These results indicated that both treatments had significant inhibitory effects on chestnut seed germination and early root growth.

Figure 1. (**A**) Effects of exogenous GABA and VGB on (**B**) chestnut seed germination and (**C,D**) primary root growth. * and ** represent significant differences between the treatments and the control (CK) at $p < 0.05$ and $p < 0.01$, respectively.

3.2. Changes to Endogenous GABA and H_2O_2 Contents

In the untreated seeds, the endogenous GABA concentration was initially 185.5 µg/g FW, but then increased to 767.3 µg/g FW at 48 h, after which it significantly decreased to 422.6 µg/g FW at 120 h (Figure 2A). However, the endogenous GABA concentrations decreased only following the VGB treatment after the 15 h imbibition (0 h). Additionally, the GABA and VGB treatments significantly decreased the endogenous GABA concentrations by over 22.1–23.5% at 48 h. Regarding the H_2O_2 contents, the CK level slightly increased. Surprisingly, the H_2O_2 contents increased considerably after the GABA and VGB treatments. After the 15 h imbibition, the H_2O_2 contents increased sharply by over 94.0% and 163.0% in response to GABA and VGB treatments, respectively, at 0 h, and remained high until 120 h (Figure 2B).

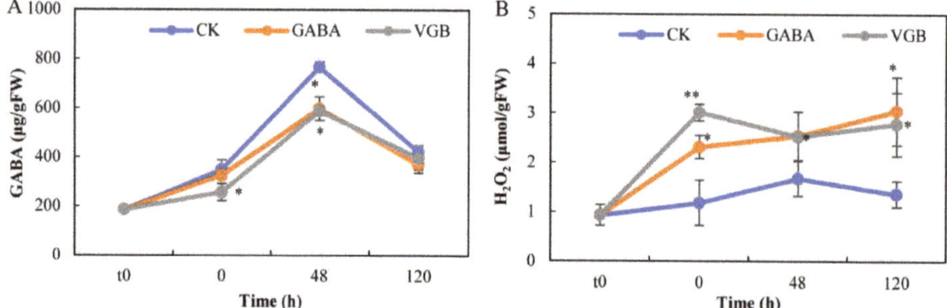

Figure 2. Effects of exogenous GABA and VGB on endogenous (**A**) GABA and (**B**) H_2O_2 contents during chestnut seed germination. * and ** represent significant differences between the treatments and control (CK) at $p < 0.05$ and $p < 0.01$, respectively. t0: time-point when the seed imbibition was initiated.

3.3. Effects of GABA and VGB on Carbon Metabolism

Compared with the CK level, the soluble sugar contents significantly increased in response to exogenous GABA only at 48 h, whereas it significantly increased at 0 and 48 h, but decreased at 120 h following the VGB treatment (Figure 3A). However, the GABA and VGB treatments did not significantly affect the starch contents (Figure 3B). Additionally, compared with the corresponding CK levels, both treatments decreased and increased the abundance of the tricarboxylic acid (TCA) cycle intermediates citrate and malate, respectively, but only at 120 h; there were no significant differences with the CK levels at all other examined time-points (Figure 4). Specifically, the citrate content decreased by over 34.2% and 65.8% (Figure 4A), and the malate content increased by over 1.8- and 5.1-fold (Figure 4B), in response to the GABA and VGB treatments, respectively. These results indicated that both treatments mainly affected the TCA cycle activity after germination.

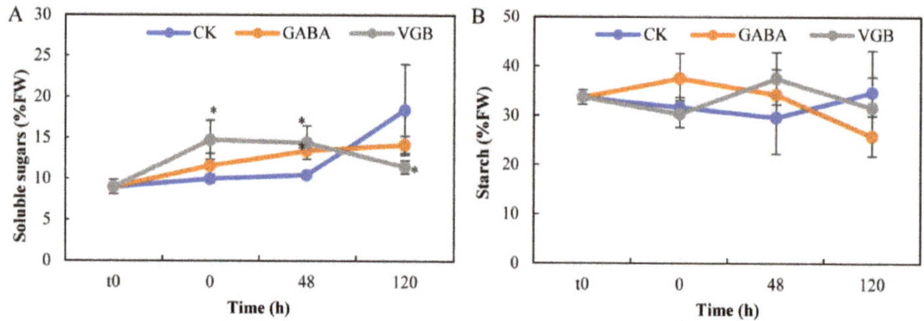

Figure 3. Effects of exogenous GABA and VGB on (**A**) the soluble sugar and (**B**) starch contents during chestnut seed germination. * represents a significant difference between the treatments and control (CK) at $p < 0.05$. t0: time-point when the seed imbibition was initiated.

3.4. Effects of GABA and VGB on Nitrogen Metabolism

Relative to the CK levels, the soluble protein contents increased in response to exogenous GABA and VGB at 120 h (Figure 5A). In contrast, only VGB increased the total amino acid content (0 h) (Figure 5B). The abundance of the following 12 free amino acids decreased after the GABA and VGB treatments at the 48 h time-point: Ser, His, Arg, Thr, Pro, Tyr, Val, Met, Lys, Ile, Phe, and Leu. At 120 h, only the Asp content was increased by both treatments, whereas the abundance of 13 free amino acids increased only in response to VGB (i.e., Ser, Glu, Gly, His, Arg, Thr, Pro, Tyr, Val, Lys, Ile, Phe, and Leu) (Figure 6). These results implied that the GABA and VGB treatments negatively regulated amino acid metabolism before germination, but after germination, the free amino acid contents were increased mainly by VGB.

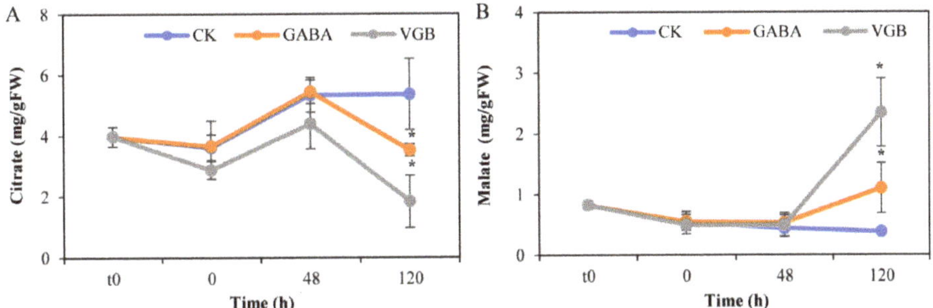

Figure 4. Effects of exogenous GABA and VGB on (**A**) the citrate and (**B**) malate contents during chestnut seed germination. * represents a significant difference between the treatments and control (CK) at $p < 0.05$. t0: time-point when the seed imbibition was initiated.

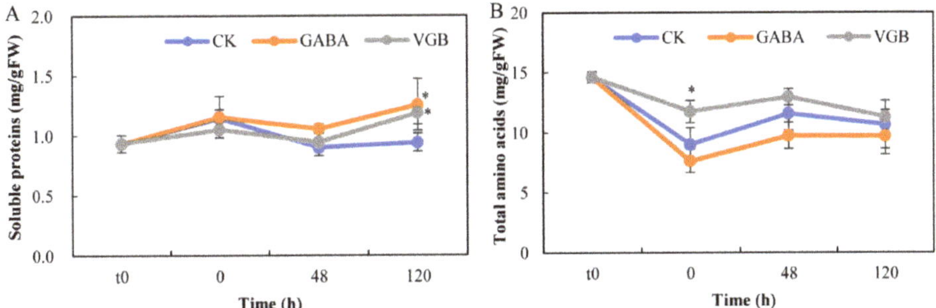

Figure 5. Effects of exogenous GABA and VGB on (**A**) soluble protein and (**B**) total amino acid contents during chestnut seed germination. * represents a significant difference between the treatments and control (CK) at $p < 0.05$. t0: time-point when the seed imbibition was initiated.

3.5. Analysis of the Physiological Response to Germination Following the GABA and VGB Treatments

We completed a PCA to explore the effects of GABA and VGB before germination (48 h) (Figure 7A) and after germination (120 h) (Figure 7B). Specifically, the physiological traits of the chestnut seeds were evaluated (Table S1). Our data revealed that PC1 and PC2 accounted for 62% and 12% of the physiological variation before germination (48 h), respectively (Figure 7A). The effects of GABA and VGB were clearly separated from the effects of CK by PC1 at 48 h. Additionally, Ser, Pro, Tyr, Val, Met, and Ile were key contributors to PC1, whereas soluble sugars and total amino acids were important factors for PC2 (Table S1). After germination (120 h), PC1 and PC2 accounted for 67% and 14% of the physiological variation, respectively (Figure 7B). The effects of the VGB treatment were clearly separated from the effects of CK; however, the effects of the GABA treatment and CK were uncovered by PC1. These findings indicated that exogenous GABA and VGB were important for inhibiting chestnut seed germination.

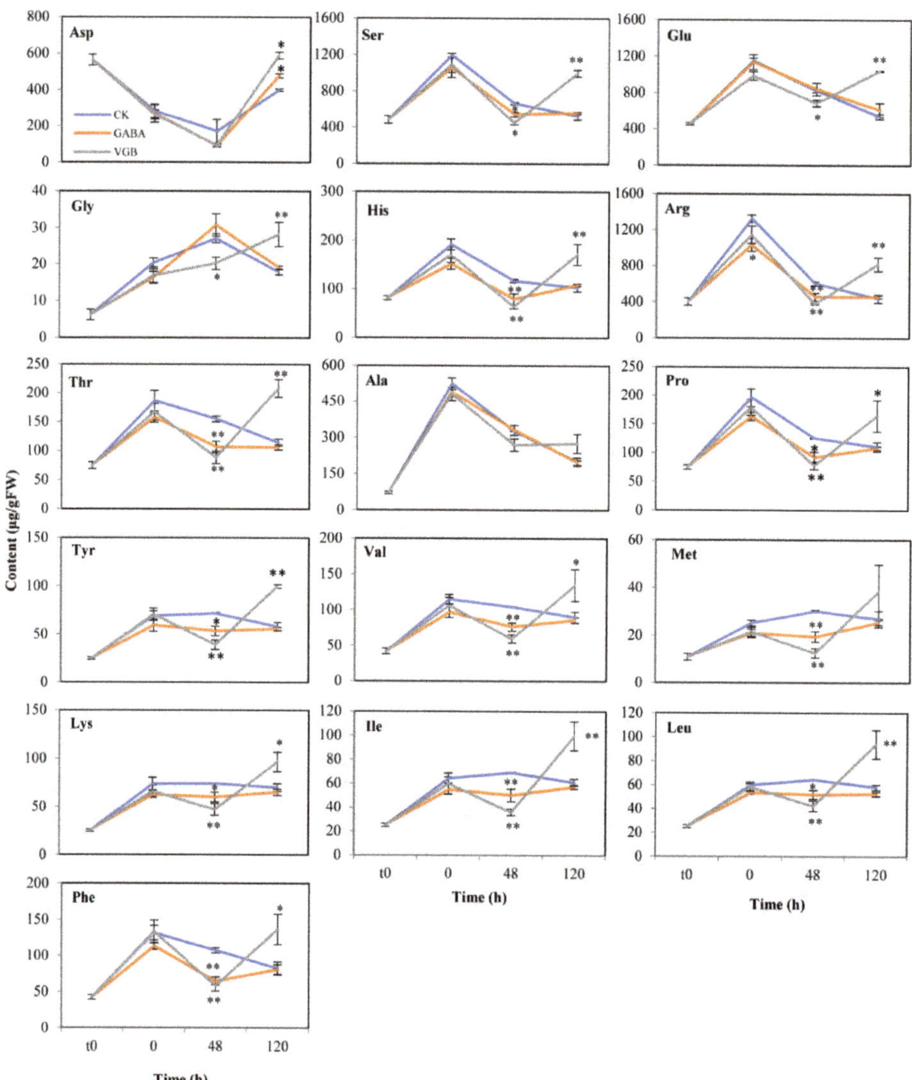

Figure 6. Effects of exogenous GABA and VGB on free amino acid contents during chestnut seed germination. * and ** represent significant differences between the treatments and control (CK) at $p < 0.05$ and $p < 0.01$, respectively. t0: time-point when the seed imbibition was initiated.

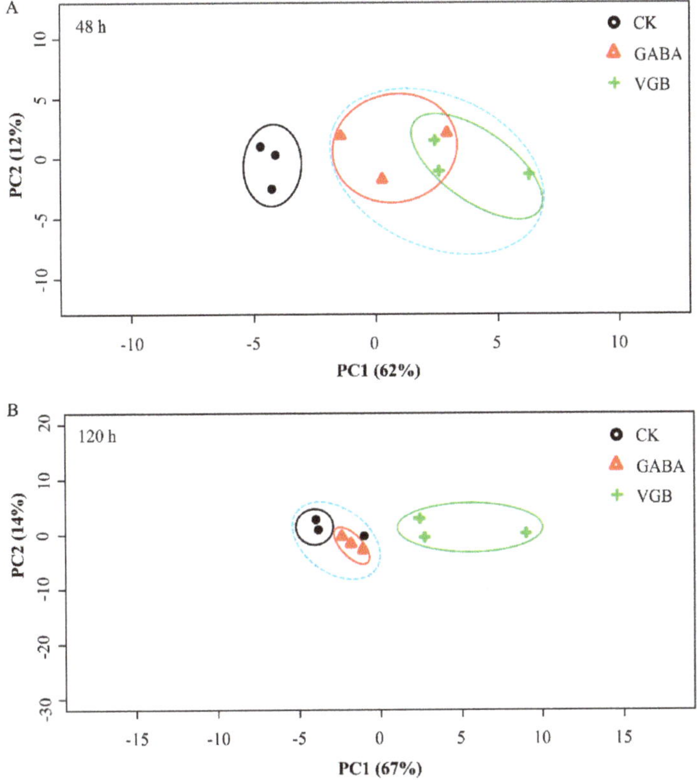

Figure 7. Principal component analysis of the effects of exogenous GABA and VGB on physiological parameters at 48 (**A**) and 120 h (**B**).

4. Discussion

Chestnut seeds are widely considered to be healthy for humans [45]. Earlier reports regarding the substantial accumulation of GABA in chestnut seeds [7,36] provided evidence of the health benefits of chestnuts, similar to other GABA-enriched functional foods [9,10]. Additionally, seed germination is closely associated with the shelf-life of chestnuts and with the cultivation of rootstock seedlings [30]. Therefore, it is worth considering the high GABA level regarding its effects on the storage and germination of chestnut seeds.

Recent genetic and physiological studies have implied that GABA is involved in responses to abiotic stresses [46,47], as well as developmental processes, including pollen tube growth [48], primary/adventitious root growth [37,49], and seed germination [20,21]. The recent identification of a GABA receptor, aluminum-activated malate transporter (ALMT) [50], indicates that GABA is a signaling molecule and not just a metabolite [51]. Consequently, GABA functions should be more comprehensively characterized.

Previous studies revealed that endogenous GABA concentrations increase during germination [16,17,52]. In contrast, in the current study, we observed that germination was negatively related to endogenous GABA concentrations at the 48 and 120 h time-points following the GABA and VGB treatments (Figures 1A and 2A). Our results were similar to those of an earlier investigation on high lysine maize seeds [13]. We also detected a more than four-fold increase in GABA concentrations before germination (Figure 2A), which might result from the decrease in Glu content at 48 h (Figure 6) because Glu acts as the direct precursor of GABA production [8,51]. This further indicated that chestnut seeds may be useful as a GABA-enriched functional food by short-term germination induction. However, the decrease in GABA after 48 h may be ascribed to the requirement of much more Glu for protein synthesis or rapid degradation of GABA back to the TCA cycle during primary root growth [8,51]. However, the application of 10 mM GABA inhibited chestnut seed germination and early primary root growth (Figure 1), which was inconsistent with the results of earlier investigations on barley [20] and *H. ammodendron* [21] seeds. Moreover, blocking GABA degradation with 100 μM VGB also had an inhibitory effect (Figure 1). We recently confirmed VGB to be detrimental to adventitious root growth [37]. Thus, the endogenous GABA may play a specific role in the germination of chestnut seeds and may be useful for improving chestnut seed storage during winter.

Generally, exogenous GABA promotes germination [20,21] and enhances GABA absorption in orthodox seeds [20]. Unexpectedly, our analysis of recalcitrant chestnut seeds uncovered a transient decrease in endogenous GABA concentrations at 48 h following the GABA and VGB treatments (Figure 2A), which was inconsistent with the data generated during our recent investigation of GABA- and VGB-treated poplar stem fragments [37]. Previous studies proved that embryo axes and cotyledons exhibited contrasting responses to desiccation in *Castanea sativa* seeds [27], and the embryonic axes of dormant seeds were maintained in a state of metabolic readiness under optimal conditions [29]. Accordingly, it is possible that the results of the current study were due to the embryos being highly sensitive to the imbibition of exogenous GABA and VGB, resulting in significant increases in endogenous GABA concentrations, relative to the levels of cotyledon with high moisture. However, the whole seed kernels used for measurements may have obscured the final increased GABA concentrations in the embryos. Additionally, GABA provides the carbon skeletons in the TCA cycle [8], which contributes to seed germination [20]. In this study, the GABA and VGB treatments altered the metabolism of soluble sugars, organic acids, and amino acids.

Carbon and nitrogen metabolites, including those mentioned above, are significantly associated with germination and seedling establishment [13], during which GABA critically affects carbon and nitrogen metabolism [8]. During seed germination, as the hormone GA's role [53], exogenous GABA may promote starch hydrolysis to produce soluble sugars by stimulating α-amylase activity [20], but in the current study, there were no significant changes in starch contents following the GABA and VGB treatments (Figure 3B). This may have resulted in the decrease in soluble sugar contents at 120 h (Figure 3A), with the resulting lack of sufficient energy leading to the inhibition of early primary root growth (Figure 1). However, we speculated that the increase in soluble sugar levels at 48 h may have resulted from the lipid breakdown occurring in germinating seeds [53]. We observed that both treatments induced the considerable accumulation of H_2O_2 (Figure 2B), which can be mainly produced by fatty acid β-oxidation during germination [53], contributing to the inhibition of chestnut seed germination.

The TCA cycle activity is closely associated with seed germination [13], wherein succinate is also the final product of GABA degradation [8]. However, the succinate contents could not be detected in this study. Previous research demonstrated that increasing lysine levels in *Arabidopsis thaliana* seeds resulted in delayed germination, which was accompanied by a significant decrease in the levels of TCA cycle metabolites, such as citrate, malate, and succinate [13]. Hence, the observed decrease in citrate contents at 120 h (Figure 4A) may adversely affect germination and early primary root growth in chestnut seeds because of the associated lack of sufficient energy. Unlike the study by Angelovici et al. [13], we detected an increase in malate contents induced by GABA and VGB treatments (Figure 4B). This increase may block the germination and early primary root establishment of chestnut seeds, which is supported by a recent report [54], which proved that rapidly germinating seeds have low malate levels. Because malate is a key intermediate of the TCA cycle, we speculated that the accumulation of malate may affect the efficient mobilization of storage compounds to supply energy for germination and early seedling development. However, the results of our recent study implied that malate interacting with GABA can delay poplar AR formation [37], possibly because GABA can negatively modulate ALMT via malate [50,55,56]. Thus, we considered that changes to the metabolic status of malate and GABA led to physiological responses, such as the delayed seed germination and inhibited early primary root growth observed in this study, through modulated ALMT activities.

In many plant species, most amino acids accumulate during seed germination [12]. For example, the aspartic acid family of amino acids contributes to the onset of autotrophic growth-associated processes during germination [13]. An exogenous nitric oxide donor (*S*-nitroso-*N*-acetyl-D,L-penicillamine) can enhance the germination of Kabuli chickpea seeds, which coincides with an increase in amino acid levels [57]. Exogenous H_2O_2 also promotes the germination of eggplant seeds, accompanied by enhanced amino acid biosynthesis and protein expression [58]. Therefore, the observed decrease in most of the amino acid levels at the 48 h time-point following the GABA and VGB treatments might contribute to the inhibition of chestnut seed germination (Figure 6). At 120 h, however, only the VGB treatment induced a considerable increase in the accumulation of 10 amino acids (Figure 6), which were reportedly negatively associated with root growth [13,37,58–60]. Thus, VGB might have a specific role in inhibiting early primary root growth. However, the GABA treatment did not adversely affect the roots, which was consistent with the findings of a previous study involving *Brassica napus* seedlings [60], but it inhibited chestnut seed germination. Furthermore, soluble protein contents reportedly decreased during the seed germination of six grass species [61], but they increased significantly when the primary root growth was inhibited at 120 h after both treatments in the current study (Figure 5). Therefore, the two treatments appeared to negatively influence early primary root growth in chestnut seeds.

5. Conclusions

The importance of GABA for human health and plant development has been confirmed. In this study, high GABA levels were detected in seeds before germination, implying GABA not only could influence chestnut seed germination, but also could act as a potential compound of functional chestnut food. This should be examined in greater detail in future studies. The application of exogenous GABA and VGB inhibited chestnut seed germination and early primary root growth, possibly by altering the balance between carbon and nitrogen metabolism, especially the free amino acid contents before germination (Figure 8). The data presented herein suggested that changes to the endogenous GABA levels in chestnut seeds might adversely affect germination. This insight may be relevant for improving the storage of chestnut and other recalcitrant seeds over winter.

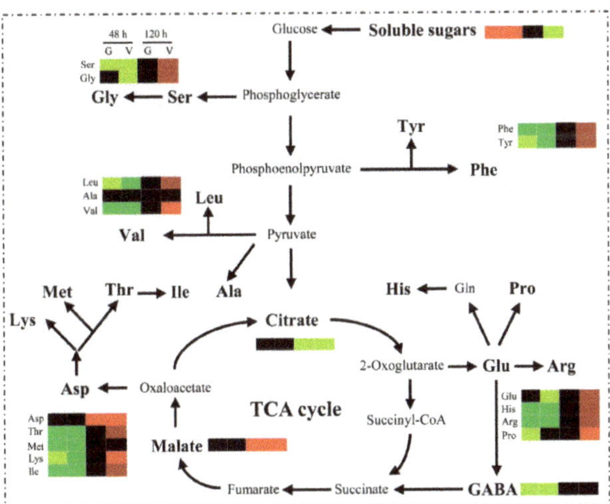

Figure 8. Model of the effects of exogenous GABA and VGB on primary carbon and nitrogen metabolism during chestnut seed germination at 48 and 120 h. Red: increase; green: decrease; black: no significant change; G: GABA vs. CK; V: VGB vs. CK.

Supplementary Materials: The following are available online at http://www.mdpi.com/2223-7747/9/4/449/s1, Table S1: Data of PCA analysis of exogenous GABA and VGB on the physiological effects during chestnut seed germination.

Author Contributions: Manuscript draft: S.S., C.D., and W.C.; experiment: C.D. and W.C.; analysis: C.D., W.C., J.S., G.W., Y.W., J.Z., J.J., and D.Y.; conception and supervision: S.S. and Z.J. All authors have read and agreed to the published version of the manuscript.

Funding: This work was supported by grants from the Project of Science and Technology Development Center of State Forestry and grassland administration (KJZXSA2019045); the National Natural Science Foundation of China (31971627); the Project of Key Laboratory of Tree Breeding and Cultivation of State Forestry and Grassland Administration (ZDRIF201712); the Science & Technology Research and Development Program of Hebei Province (16226312D).

Acknowledgments: We thank Liwen Bianji, Edanz Editing China (www.liwenbianji.cn/ac), for editing the English text of a draft of this manuscript.

Conflicts of Interest: The authors declare no conflict of interest.

References

1. LaBonte, N.R.; Zhao, P.; Woeste, K. Signatures of selection in the genomes of Chinese chestnut (*Castanea mollissima* Blume): The roots of nut tree domestication. *Front. Plant Sci.* **2018**, *9*, 810. [CrossRef] [PubMed]

2. Xing, Y.; Liu, Y.; Zhang, Q.; Nie, X.; Sun, Y.; Zhang, Z.; Li, H.; Fang, K.; Wang, G.; Huang, H.; et al. Hybrid de novo genome assembly of Chinese chestnut (*Castanea mollissima*). *GigaScience* **2019**, *8*, pii:giz112. [CrossRef] [PubMed]

3. Staton, M.; Zhebentyayeva, T.; Olukolu, B.; Fang, G.C.; Nelson, D.; Carlson, J.E.; Abbott, A.G. Substantial genome synteny preservation among woody angiosperm species: Comparative genomics of Chinese chestnut (*Castanea mollissima*) and plant reference genomes. *BMC Genomics* **2015**, *16*, 744. [CrossRef]

4. Rigling, D.; Prospero, S. *Cryphonectria parasitica*, the causal agent of chestnut blight: Invasion history, population biology and disease control. *Mol. Plant Pathol.* **2018**, *19*, 7–20. [CrossRef] [PubMed]

5. De Vasconcelos, M.C.; Bennett, R.N.; Rosa, E.A.; Ferreira-Cardoso, J.V. Composition of European chestnut (*Castanea sativa* Mill.) and association with health effects: Fresh and processed products. *J. Sci. Food Agric.* **2010**, *90*, 1578–1589. [CrossRef] [PubMed]

6. Zhang, L.; Lin, Q.; Feng, Y.; Fan, X.; Zou, F.; Yuan, D.Y.; Zeng, X.; Cao, H. Transcriptomic identification and expression of starch and sucrose metabolism genes in the seeds of Chinese chestnut (*Castanea mollissima*). *J. Agric. Food Chem.* **2015**, *63*, 929–942. [CrossRef] [PubMed]

7. Desmaison, A.M.; Tixier, M. Acides amines libres de chataignes provenant de *Castanea sativa* Mill., *Castanea crenata* Sieb et Zucc.; *Castanea mollissima* Bl. et d'hybrides: *Castanea crenata* × *Castanea sativa*. *Ann. Pharm. Fr.* **1984**, *42*, 353–357.

8. Michaeli, S.; Fromm, H. Closing the loop on the GABA shunt in plants: Are GABA metabolism and signaling entwined? *Front. Plant Sci.* **2015**, *6*, 419. [CrossRef]

9. Wu, Q.; Shah, N.P. High γ-aminobutyric acid production from lactic acid bacteria: Emphasis on *Lactobacillus brevis* as a functional dairy starter. *Crit. Rev. Food Sci. Nutr.* **2017**, *57*, 3661–3672. [CrossRef]

10. Wang, P.; Liu, K.; Gu, Z.; Yang, R. Enhanced γ-aminobutyric acid accumulation, alleviated componential deterioration and technofunctionality loss of germinated wheat by hypoxia stress. *Food Chem.* **2018**, *269*, 473–479. [CrossRef]

11. Greggains, V.; Finch-Savage, W.; Quick, W.; Atherton, N. Putative desiccation tolerance mechanisms in orthodox and recalcitrant seeds of the genus *Acer*. *Seed Sci. Res.* **2000**, *10*, 317–332. [CrossRef]

12. Fait, A.; Angelovici, R.; Less, H.; Ohad, I.; Urbanczyk-Wochniak, E.; Fernie, A.R.; Galili, G. *Arabidopsis* seed development and germination is associated with temporally distinct metabolic switches. *Plant Physiol.* **2006**, *142*, 839–854. [CrossRef] [PubMed]

13. Angelovici, R.; Fait, A.; Fernie, A.R.; Galili, G. A seed high-lysine trait is negatively associated with the TCA cycle and slows down *Arabidopsis* seed germination. *New Phytol.* **2011**, *189*, 148–159. [CrossRef] [PubMed]

14. Peter, K.; Georg, H.; Herbert, W.; Michael, R. Changes of folates, dietary fiber, and proteins in wheat as affected by germination. *J. Agric. Food Chem.* **2007**, *55*, 4678–4683.

15. Joshi, P.; Varma, K. Effect of germination and dehulling on the nutritive value of soybean. *Nutr. Food Sci.* **2016**, *46*, 595–603. [CrossRef]

16. Sharma, S.; Saxena, D.C.; Riar, C.S. Analysing the effect of germination on phenolics, dietary fibres, minerals and γ-aminobutyric acid contents of barnyard millet (*Echinochloa frumentaceae*). *Food Biosci.* **2016**, *13*, 60–68. [CrossRef]

17. Kim, M.; Kwak, H.; Kim, S. Effects of germination on protein, γ-aminobutyric acid, phenolic acids, and antioxidant capacity in wheat. *Molecules* **2018**, *23*, 2244. [CrossRef]

18. AL-Quraan, N.A.; AL-Ajlouni, Z.I.; Obedat, D.I. The GABA shunt pathway in germinating seeds of wheat (*Triticum aestivum* L.) and barley (*Hordeum vulgare* L.) under salt stress. *Seed Sci. Res.* **2019**, *29*, 1–11. [CrossRef]

19. Zhao, G.C.; Xie, M.X.; Wang, Y.C.; Li, J.Y. Molecular mechanisms underlying γ-aminobutyric acid (GABA) accumulation in giant embryo rice seeds. *J. Agric. Food Chem.* **2017**, *65*, 4883–4889. [CrossRef]

20. Sheng, Y.; Xiao, H.; Guo, C.; Wu, H.; Wang, X. Effects of exogenous gamma-aminobutyric acid on α-amylase activity in the aleurone of barley seeds. *Plant Physiol. Biochem.* **2018**, *127*, 39. [CrossRef]

21. Shi, Z.; Shi, S.Q.; Zhong, C.F.; Yao, H.J.; Gao, R.F. Effect on germination and respiration of *Haloxylon ammodendron* seeds by the application of exogenous Glu and GABA. *Acta Bot. Boreal. Occident. Sin.* **2008**, *28*, 1455–1460.

22. Shi, S.Q.; Zheng, S.; Jiang, Z.P.; Qi, L.W.; Sun, X.M.; Li, C.X.; Liu, J.F.; Xiao, W.F.; Zhang, S.G. Effects of exogenous GABA on gene expression of *Caragana intermedia* roots under NaCl stress: Regulatory roles for H_2O_2 and ethylene production. *Plant Cell Environ.* **2010**, *33*, 149–162. [CrossRef] [PubMed]

23. Ji, J.; Yue, J.; Xie, T.; Chen, W.; Du, C.; Chang, E.; Chen, L.; Jiang, Z.; Shi, S. Roles of γ-aminobutyric acid on salinity-responsive genes at transcriptomic level in poplar: Involving in abscisic acid and ethylene-signalling pathways. *Planta* **2018**, *248*, 1–16. [CrossRef] [PubMed]

24. Cheng, B.; Li, Z.; Liang, L.; Cao, Y.; Zeng, W.; Zhang, X.; Ma, X.; Huang, L.; Nie, G.; Liu, W. The γ-aminobutyric acid (GABA) alleviates salt stress damage during seeds germination of white clover associated with Na^+/K^+ transportation, dehydrins accumulation, and stress-related genes expression in white clover. *Int. J. Mol. Sci.* **2018**, *19*, 2520. [CrossRef]

25. Ma, Y.; Wang, P.; Wang, M.; Sun, M.; Gu, Z.; Yang, R. GABA mediates phenolic compounds accumulation and the antioxidant system enhancement in germinated hulless barley under NaCl stress. *Food Chem.* **2019**, *270*, 593–601. [CrossRef]

26. Han, S.; Nan, Y.; Qu, W.; He, Y.; Ban, Q.; Lv, Y.; Rao, J. Exogenous γ-aminobutyric acid (GABA) treatment contributes to regulation of malate metabolism and ethylene synthesis in apple fruit during storage. *J. Agric. Food Chem.* **2018**, *66*, 13473–13482. [CrossRef]

27. Leprince, O.; Buitink, J.; Hoekstra, F.A. Axes and cotyledons of recalcitrant seeds of *Castanea sativa* Mill. Exhibit contrasting responses of respiration to drying in relation to desiccation sensitivity. *J. Exp. Bot.* **1999**, *388*, 1515–1524. [CrossRef]

28. Wang, G.; Liang, L.; Zong, Y. Controlling of dormancy and sprouting of stored chestnut by means of temperature. *Sci. Silvae Sin.* **1999**, *35*, 29–33.

29. Obroucheva, N.; Sinkevich, I.; Lityagina, S. Physiological aspects of seed recalcitrance: A case study on the tree *Aesculus hippocastanum*. *Tree Physiol.* **2016**, *36*, 1127–1150. [CrossRef]

30. Karadeniz, T.; Zeki Bostan, S.; Islam, A. The effects of different applications on seed germination in chestnut. *Acta Hort.* **2013**, *981*, 461–463. [CrossRef]

31. Soylu, A.; Eris, A.; Özgür, M.; Dalkiliç, Z. Researches on the rootstock potentiality of chestnut types (*Castanea sativa* Mill.) grown in Marmara region. *Acta Hort.* **1999**, *494*, 213–222. [CrossRef]

32. Soylu, A.; Serdar, U. Rootstock selection on chestnut (*Castanea sativa* Mill.) in the middle of black sea region in Turkey. *Acta Hort.* **2000**, *538*, 483–487. [CrossRef]

33. Chen, H.; Ruan, J.; Chu, P.; Fu, W.; Liang, Z.; Li, Y.; Tong, J.; Xiao, L.; Liu, J.; Li, C.; et al. AtPER1 enhances primary seed dormancy and reduces seed germination by suppressing the ABA catabolism and GA biosynthesis in *Arabidopsis* seeds. *Plant J.* **2020**, *101*, 310–323. [CrossRef] [PubMed]

34. Xu, K.; Sun, Q.; Xiao, S. Studies on dormancy and germination of China chestnut seeds. *Chin. Agric. Sci. Bull.* **1998**, *14*, 24–25, 28.

35. Roach, T.; Beckett, R.P.; Minibayeva, F.V.; Colville, L.; Whitaker, C.; Chen, H.; Bailly, C.; Kranner, I. Extracellular superoxide production, viability and redox poise in response to desiccation in recalcitrant *Castanea sativa* seeds. *Plant Cell Environ.* **2010**, *33*, 59–75.

36. Desmaison, A.M.; Tixier, M. Amino acids content in germinating seeds and seedlings from *Castanea sativa* L. *Plant Physiol.* **1986**, *81*, 692–695. [CrossRef]

37. Xie, T.; Ji, J.; Chen, W.; Yue, J.; Du, C.; Sun, J.; Chen, L.; Jiang, Z.; Shi, S. GABA negatively regulates adventitious root development. *J. Exp. Bot.* **2020**, *71*, 1459–1474. [CrossRef]

38. Wang, B.; Guo, X.; Zhao, P.; Ruan, M.; Yu, X.; Zou, L.; Yang, Y.; Li, X.; Deng, D.; Xiao, J.; et al. Molecular diversity analysis, drought related marker-traits association mapping and discovery of excellent alleles for 100-day old plants by EST-SSRs in cassava germplasms (*Manihot esculenta* Cranz). *PLoS ONE* **2017**, *12*, e0177456. [CrossRef]

39. Yemm, E.W.; Willis, A.J. The estimation of carbohydrates in plant extracts by anthrone. *Biochem. J.* **1954**, *57*, 508–514. [CrossRef] [PubMed]

40. López-Delgado, H.; Zavaleta-Mancera, H.A.; Mora-Herrera, M.E.; Vázquez-Rivera, M.; Flores-Gutiérrez, F.X.; Scott, I.M. Hydrogen peroxide increases potato tuber and stem starch content, stem diameter, and stem lignin content. *Amer. J. Potato Res.* **2005**, *82*, 279–285. [CrossRef]

41. Chen, Y.; Fu, X.; Mei, X.; Zhou, Y.; Cheng, S.; Zeng, L.; Dong, F.; Yang, Z. Proteolysis of chloroplast proteins is responsible for accumulation of free amino acids in dark-treated tea (*Camellia sinensis*) leaves. *J. Proteomics* **2017**, *157*, 10–17. [CrossRef] [PubMed]

42. Yue, J.; Du, C.; Ji, J.; Xie, T.; Chen, W.; Chang, E.; Chen, L.; Jiang, Z.; Shi, S. Inhibition of α-ketoglutarate dehydrogenase activity affects adventitious root growth in poplar via changes in GABA shunt. *Planta* **2018**, *248*, 963–979. [CrossRef] [PubMed]

43. Luo, J.; Li, H.; Liu, T.; Polle, A.; Peng, C.; Luo, Z.B. Nitrogen metabolism of two contrasting poplar species during acclimation to limiting nitrogen availability. *J. Exp. Bot.* **2013**, *64*, 4207–4224. [CrossRef] [PubMed]

44. Zheng, H.; Zhang, X.; Ma, W.; Song, J.; Rahman, S.U.; Wang, J.; Zhang, Y. Morphological and physiological responses to cyclic drought in two contrasting genotypes of *Catalpa bungei*. *Environ. Exp. Bot.* **2017**, *138*, 77–87. [CrossRef]

45. Mujić, I.; Agayn, V.; Živković, J.; Velić, D.; Jokić, S.; Alibabić, V.; Rekić, A. Chestnuts, a "comfort" healthy food? *Acta Hort.* **2010**, *866*, 659–665. [CrossRef]

46. Bao, H.; Chen, X.; Lv, S.; Jiang, P.; Feng, J.; Fan, P.; Nie, L.; Li, Y. Virus-induced gene silencing reveals control of reactive oxygen species accumulation and salt tolerance in tomato by γ-aminobutyric acid metabolic pathway. *Plant Cell Environ.* **2015**, *38*, 600–613. [CrossRef]

47. Che-Othman, M.H.; Jacoby, R.P.; Millar, A.H.; Taylor, N.L. Wheat mitochondrial respiration shifts from the tricarboxylic acid cycle to the GABA shunt under salt stress. *New Phytol.* **2019**, *225*, 1166–1180. [CrossRef]

48. Palanivelu, R.; Brass, L.; Edlund, A.F.; Preuss, D. Pollen tube growth and guidance is regulated by *POP2*, an *Arabidopsis* gene that controls GABA levels. *Cell* **2003**, *114*, 47–59. [CrossRef]

49. Barbosa, J.M.; Singh, N.K.; Cherry, J.H.; Locy, R.D. Nitrate uptake and utilization is modulated by exogenous γ-aminobutyric acid in *Arabidopsis thaliana* seedlings. *Plant Physiol. Biochem.* **2010**, *48*, 443–450. [CrossRef]

50. Ramesh, S.A.; Tyerman, S.D.; Xu, B.; Bose, J.; Kaur, S.; Conn, V.; Domingos, P.; Ullah, S.; Wege, S.; Shabala, S.; et al. GABA signalling modulates plant growth by directly regulating the activity of plant-specific anion transporters. *Nat. Commun.* **2015**, *6*, 7879. [CrossRef]

51. Bown, A.W.; Shelp, B.J. Plant GABA: Not just a metabolite. *Trend Plant Sci.* **2016**, *21*, 811–813. [CrossRef] [PubMed]

52. Gómez-Favela, M.A.; Gutiérrez-Dorado, R.; Cuevas-Rodríguez, E.O.; Canizalez-Román, V.A.; Del Rosario León-Sicairos, C.; Milán-Carrillo, J.; Reyes-Moreno, C. Improvement of chia seeds with antioxidant activity, GABA, essential amino acids, and dietary fiber by controlled germination bioprocess. *Plant Foods Hum. Nutr.* **2017**, *72*, 345–352. [CrossRef]

53. Sun, J.; Jia, H.; Wang, P.; Zhou, T.; Wu, Y.; Liu, Z. Exogenous gibberellin weakens lipid breakdown by increasing soluble sugars levels in early germination of zanthoxylum seeds. *Plant Sci.* **2019**, *280*, 155–163. [CrossRef] [PubMed]

54. Boter, M.; Calleja-Cabrera, J.; Carrera-Castaño, G.; Wagner, G.; Hatzig, S.V.; Snowdon, R.J.; Legoahec, L.; Bianchetti, G.; Bouchereau, A.; Nesi, N.; et al. An integrative approach to analyze seed germination in *Brassica napus*. *Front. Plant Sci.* **2019**, *10*, 1342. [CrossRef]

55. Ramesh, S.A.; Tyerman, S.D.; Gilliham, M.; Xu, B. γ-Aminobutyric acid (GABA) signalling in plants. *Cell Mol. Life Sci.* **2017**, *74*, 1577–1603. [CrossRef] [PubMed]

56. Ramesh, S.A.; Kamran, M.; Sullivan, W.; Chirkova, L.; Okamoto, M.; Degryse, F.; Mclauchlin, M.; Gilliham, M.; Tyerman, S.D. Aluminium-activated malate transporters can facilitate GABA transport. *Plant Cell* **2018**, *30*, 1147–1164. [CrossRef]

57. Pandey, S.; Kumari, A.; Shree, M.; Kumar, V.; Singh, P.; Bharadwaj, C.; Loake, G.J.; Parida, S.K.; Masakapalli, S.K.; Gupta, K.J. Nitric oxide accelerates germination via the regulation of respiration in chickpea. *J. Exp. Bot.* **2019**, *70*, 4539–4555. [CrossRef]

58. Dufková, H.; Berka, M.; Luklová, M.; Rashotte, A.M.; Brzobohatý, B.; Černý, M. Eggplant germination is promoted by hydrogen peroxide and temperature in an independent but overlapping manner. *Molecules* **2019**, *24*, 4270. [CrossRef]

59. Han, R.; Khalid, M.; Juan, J.; Huang, D. Exogenous glycine inhibits root elongation and reduces nitrate-N uptake in pak choi (*Brassica campestris* ssp. *Chinensis* L.). *PLoS ONE* **2018**, *13*, e0204488. [CrossRef]

60. Deleu, C.; Faes, P.; Niogret, M.F.; Bouchereau, A. Effects of the inhibitor of the γ-aminobutyrate-transaminase, vinyl-γ-aminobutyrate, on development and nitrogen metabolism in *Brassica napus* seedlings. *Plant Physiol. Biochem.* **2013**, *64*, 60–69. [CrossRef]

61. Zhao, M.; Zhang, H.; Yan, H.; Qiu, L.; Baskin, C.C. Mobilization and role of starch, protein, and fat reserves during seed germination of six wild grassland species. *Front. Plant Sci.* **2018**, *9*, 234. [CrossRef] [PubMed]

Review

Auxin: Hormonal Signal Required for Seed Development and Dormancy

Angel J. Matilla

Departamento de Biología Funcional (Área Fisiología Vegetal), Facultad de Farmacia, Universidad de Santiago de Compostela, 15782 Santiago de Compostela, Spain; angeljesus.matilla@usc.es; Tel.: +34-981-563-100

Received: 28 April 2020; Accepted: 27 May 2020; Published: 1 June 2020

Abstract: The production of viable seeds is a key event in the life cycle of higher plants. Historically, abscisic acid (ABA) and gibberellin (GAs) were considered the main hormones that regulate seed formation. However, auxin has recently emerged as an essential player that modulates, in conjunction with ABA, different cellular processes involved in seed development as well as the induction, regulation and maintenance of primary dormancy (PD). This review examines and discusses the key role of auxin as a signaling molecule that coordinates seed life. The cellular machinery involved in the synthesis and transport of auxin, as well as their cellular and tissue compartmentalization, is crucial for the development of the endosperm and seed-coat. Thus, auxin is an essential compound involved in integuments development, and its transport from endosperm is regulated by AGAMOUS-LIKE62 (AGL62) whose transcript is specifically expressed in the endosperm. In addition, recent biochemical and genetic evidence supports the involvement of auxins in PD. In this process, the participation of the transcriptional regulator ABA INSENSITIVE3 (ABI3) is critical, revealing a cross-talk between auxin and ABA signaling. Future experimental aimed at advancing knowledge of the role of auxins in seed development and PD are also discussed.

Keywords: ABA; primary dormancy; ABI3; auxin; YUC; PIN; ARF; endosperm; integuments; AGL62; PRC2

1. Introduction

The evolutionary success of higher plants consists of their ability to produce seeds, units responsible for reproduction, dispersal and survival [1]. Synchronized coordination between hormone signaling networks and environmental cues are being required to control these processes. The viable seed is an entity that originates at the end of the development program progression from the fertilized egg and it is constituted of three genetically different compartments [2,3]. That is filial endosperm (3n) and embryo (2n) on the one hand, and maternal seed-coat (2n) on the other [4,5]. The Angiosperm seed development initiates when the paternal and maternal gametes fuse to create the diploid embryo and the triploid endosperm. In most higher plants, the endosperm initially develops as a syncytium, in which nuclear divisions are not followed by cytokinesis. After a specific number of caryokinesis, the endosperm becomes cellularized. However, the mechanism that regulates the transition to cellularization, a critical process in seed development, remains unknown [6]. The endosperm is usually consumed in the dicots during seed development, while it is retained in mature seeds of monocots. However, the developmental process of endosperm is quite highly conserved in plants. The embryo arrest and seed lethality are produced when endosperm cellularization is impaired. Interestingly, the auxin levels need to be tightly controlled to allow the endosperm to cellularize [7]. On the other hand, seed development relies on a strong interdependent control between three respective compartments that constitute it [3,8–10]. Therefore, it is not surprising that all molecular events involved in zygotic embryogenesis are tightly coordinated at the genetic and hormonal levels [9,11,12]. The analysis of zygotic embryogenesis has been basically carried out by the characterization of mutants [13–15]. Once the seed tissues are completely differentiated, the embryogenic phase ends and begins the maturation phase in which storage compounds (i.e., proteins and lipids) accumulate in the endosperm (monocots) or in cotyledons (eudicots). Throughout maturation, desiccation tolerance is acquired and programmed cell death occurs; finally, the primary seed dormancy is triggered preventing vivipary [1,16–18].

The phytohormone abscisic acid (ABA) regulates multiple physiological processes, including seed maturation, embryo morphogenesis and desiccation, stomatal movements, and synthesis of stress proteins and metabolites [19,20]. ABA is the only hormone known to induce, regulate, and maintain the primary seed dormancy. Thus, seeds of ABA-deficient mutants germinate faster than the wild-type, and transgenic plants constitutively expressing the ABA biosynthesis gene maintain deep seed dormancy [1,21]. During seed development, ABA is produced in all seed compartments, as suggested by the spatiotemporal expression of its biosynthesis genes [22,23]. ABA synthesized in the endosperm and then transported to the embryo is involved in the induction of seed dormancy [22–25]. Likewise, ABA shows an accumulation pattern complementary to the gibberellin (GAs), being the main hormone that inhibits all the processes induced by them [26–28]. Regarding ABA signaling, ABA receptors PYRABACTIN RESISTANT/PYRABACTIN RESISTANT-LIKE/REGULATORY COMPONENT OF ABA RECEPTORs (PYR/PYL/RCAR) bind to ABA to remove the repression by PP2Cs (protein phosphatase 2C) of ABA responses [29]. ABA receptors constitute a 14-member family [30]. The PP2C Arabidopsis cluster includes nine members (i.e., ABI1, ABI2, HAB1, HAB2, AHG1, AHG3/AtPP2CA, HAI1, HAI2 and HAI3) which are negative regulators of early ABA signaling [17,31–33]. Removal of PP2C repression allows downstream signaling via OST1/SnRK2.6/SnRK2E SNF1-related protein kinases 2 (SnRK2) [32–35] (Figure 1). The Arabidopsis genome contains 10 members of *SnRK2*; among them, *SRK2D/SnRK2.2*, *SRK2E/OST1/SnRK2.6* and *SRK2I/SnRK2.3* are essential for ABA responses [35]. The phosphorylation of proteins plays a key role in this ABA signaling pathway [36]. Parallel to what is indicated, ABA is also involved in the regulation and the mechanism of action of DELAY OF GERMINATION-1 (DOG1) a heme-binding protein and master regulator of primary dormancy (PD) that acts in concert with ABA to delay germination [1,11,37–39]. Thus, PD and germination are regulated by ABA signaling through a DOG1-AHG1 interaction, acting in parallel with PYL/RCAR ABA receptor-dependent regulation [39]. *DOG1* has been identified in A. thaliana as one of the major regulators of natural PD in conjunction with ABA [40,41]. DOG1 is mainly expressed in seeds, in particular in the vascular tissues of the developing embryo [42]. Besides, it was demonstrated that DOG1 function is not strictly limited to seed dormancy, but that it is required for other aspects of seed maturation, in part by attenuating with ABA and ethylene signaling components [37,43]. Recently, it was demonstrated that ethylene signaling controls seed dormancy via DOG1 regulation [37,44].

On the other hand, a series of evidence clearly relates ABA to the mechanism and mode of action of auxins. That is, it seems doubtless that ABA interacts with auxin to regulate various aspects of plant growth and development [45]. Therefore, some evolutionary crosstalk must occur between both plant hormones. However, the study on the participation of auxin in the final part of seed development (e.g., induction, maintenance and loss of PD) is not developed enough yet. The auxin is a signaling molecule that is present across all domains of life, including algal, moss, liverworts, lycophytes and microorganisms [46–48]. Tryptophan (L-Trp) serves as a common precursor for IAA synthesis in plants and auxin producing bacteria (Figure 1). The auxin is synthesized, stored, and inactivated by a multitude of parallel pathways that are all tightly regulated [48]. Regarding the seed, it is now widely accepted that auxin biosynthesis is required for an array of seed developmental processes (e.g., zygotic embryogenesis and endosperm development, among others) [9,49]. High levels of free-auxins and metabolites found during both early (i.e., cell division and expansion) and last phases of seed development (e.g., endosperm cellularization) suggest that auxin has an essential signaling role [9,50,51]. Recent studies have shown that auxin possesses positive effects on seed dormancy, being in conjunction with ABA the second hormone that induces seed dormancy. Thus, Liu et al. (2013) demonstrate, at the molecular level, a role for auxin in seed dormancy through stimulation of ABA signaling, identifying auxin as a promoter of seed dormancy [52]. On the other hand, the auxin also affects seed germination by altering the ABA/GAs ratio [53]. Until now, the role of phytohormones in zygotic embryogenesis mainly refers to the study of eudicots such as Arabidopsis [54]. In order to generate the appropriate response, the auxin polar transport causes its accumulation in specific cellular places. However, very little is known about auxin biosynthesis and homeostasis, polar auxin transport, and response during early embryogenesis in monocots. Interestingly, some of these features involve auxins that seem to be conserved in both monocots and dicots seeds [55,56]. Auxin is perceived by a transient co-receptor complex consisting of a TRANSPORT INHIBITOR1/AUXIN-SIGNALING F-BOX (TIR1/AFB) binding proteins (i.e., a family composed of six members in Arabidopsis) and a transcriptional repressor Aux/IAA protein whose proteosome degradation is crucial for auxin action [48]. Regarding auxin signaling, auxin-inducible genes (AIG) have AUXIN RESPONSE ELEMENTS (AREs) in their promoters, which are bound by dimers of the AUXIN RESPONSE FACTOR (ARF) TFs [57]. ARFs are TFs that regulate the expression of auxin-responsive genes [57]. In the absence of auxin or in the presence of low levels, AIG expression is prevented by the recruitment of Aux/IAA transcriptional repressors to the promoters via their interaction with the ARFs [58]. ARFs and Aux/IAAs are encoded in Arabidopsis by large gene families with 23 and 29 members, respectively. In the presence of high levels of auxin, AUX/IAA becomes ubiquitinated by the action of the multi-protein E3 ubiquitin ligase complex (SCF^{TIR1}) and are broken down by the proteosome complex. ARF can then function, often forming ARF-ARF dimmers that allow the AIGs transcription (Figure 1). The auxin-signaling pathway seems to be conserved in land plants [48].

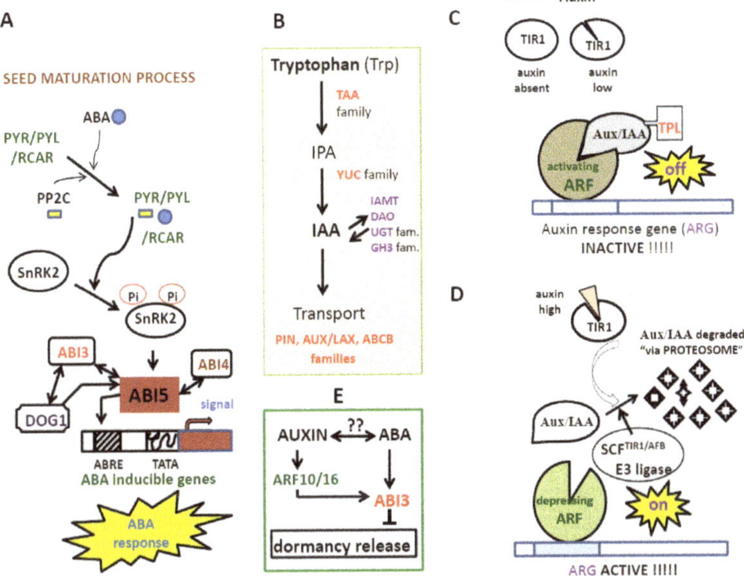

Figure 1. (**A**) In the presence of ABA, the ABA receptors PYR/PYL/RCAR form a complex with PP2C, and this inhibits the phosphatase activity of PP2C and thereby activate SnRK2. The activated SnRK2 subsequently turns on ABRE-binding protein/ABRE-binding factor (AREB/ABF) transcription factors (TFs), which in turn activates the transcription of ABA-responsive genes. Among AREB/ABF TFs, ABA insensitive 5 (ABI5), a member of the basic leucine zipper transcription factor family, plays a central role in regulating ABA-responsive genes in seeds. ABI4 and ABI3, AP2-type and B3-type TFs, respectively, have been reported to function together ABI5 to induce the expression of ABA-responsive genes, and thereby regulate seed dormancy and germination. (**B**) Auxin synthesis and transport involving TAA and YUC enzymes, and PIN, AUX/LAX and ABCB proteins. (**C**) When auxin levels are low, AUX/IAAs prevent ARF regulatory action on auxin-responsive genes. (**D**) If the cellular auxin level is high, auxin promotes interaction between TIR1/AFB and Aux/IAA proteins, resulting in degradation of the Aux/IAAs and the release of ARF repression. (**E**) Hypothesis of interaction between auxin-ABA to inhibit the primary seed dormancy release with ABI3 as a trigger.

A striking aspect of the above lies in the participation of auxin as a key hormone, in conjunction with ABA, in the regulation of specific phases of seed life. That is why this review provides the progress made in recent years on the contribution of auxin in the fertilization process and zygotic and post-zygotic embryogenesis phases. Given the recent demonstration of auxin involvement in the seed PD process, this update also considers the events that have led to this outstanding discovery.

2. Key Role of Auxin in Zygotic Embryogenesis

2.1. Spatiotemporal Auxin Production during Early Embryogenesis

Seed development encompasses a set of morphological, physiological, and biochemical changes and can be divided into three main phases: embryogenesis (including cell division and expansion, and the beginning of endosperm and embryo development), seed maturation and desiccation [54,59]. Notably, several patterning processes are controlled by auxins. The organs and tissues involved in embryogenesis require, among other control mechanisms, precise coordination between cell division and cell differentiation. Embryogenesis is initiated by the zygote polarization (i.e., embryo proper and suspensor, both symplastically connected). That is, the first step of embryonic patterning is the establishment of the apical-basal axis, in which asymmetric distribution of auxin mediated by PIN proteins plays a major role. In the zygote of maize and rice, genome activation occurs shortly after fertilization (i.e., 12 h after pollination) [60]. It is noteworthy in maize that this activation coincides with a remarkable up-regulation of a number of auxin-related genes; namely, those involved in auxin biosynthesis and signaling [60]. Although the three main organs that constitute the seed (i.e., seed-coat, endosperm and embryo) exhibit different morphology and functions, they must coordinate their growth in order to achieve the seed viability [61]. Therefore, phytohormones (i.e., auxin, cytokinins-CKs- and GAs) play key roles in the commissioning and maintenance of this strict regulation promoted by developmental program [56]. The evidenced presence of auxin during all seed development phases suggests that this hormone signaling has a consistent and key role throughout seed formation [7,9,55]. Thus, mutants deficient in auxin biosynthesis, transport, and response are defective in embryogenesis [55]. Moreover, it has been shown that auxin regulation of seed development is concentration-dependent [61]. At present, it is well known that auxin has a critical task for the ovule fertilization, subsequent embryogenesis, and determination of the young embryo polarity [54,55,62,63], among other functions. Likewise, seed-produced auxin is of importance for development and growth, and coordination of the three seed constituent organs [56]. As it happens in other plant organs, the auxin distribution in the seed depends on its polar transport [48,49,63]. Thus, auxin plays a critical role in plant growth and development by forming local concentration gradients [61]. Therefore, local auxin synthesis and metabolism, intercellular transport (i.e., AUXIN1/LIKE-AUX1 family of auxin influx carriers (AUX/LAX), PIN-FORMED family of auxin efflux carriers (PIN; with eight members in Arabidopsis and four expressed during embryogenesis), and some members of P-GLYCOPROTEIN/ATP-BINDING CASSETTE B4 (ABCB/PGP) family carriers) and auxin signaling (Figure 1), acting all in connection, will determine the gradient of auxins and tissue patterning [49,56,64–66] Cellular localization of these carriers is indicative of the auxin flow direction, creating thus morphogenic auxin gradients. That is, auxin has a vital role in determining embryo identity and structure (i.e., embryo axis formation and apical-basal pattern formation during embryogenesis).

It seems now indubitable that the biosynthesis of IAA, the main auxin of plants, takes place from L-TrP and indole-3-pyruvic acid (IPA) as the only intermediary [62]. The L-TrP aminotransferases of the Arabidopsis family (TAA; also known as WEAK ETHYLENE INSENSITIVE8-WEI8) and YUCCA (YUC)-type family of flavin-containing monooxygenases seems to act coordinately to control the IAA biosynthesis in Arabidopsis. Recently, the conservation and diversification of TAA and YUCCA functions were highlighted [62]. There are 5 *TAA* and 11 *YUC* gene members identified in the genome of Arabidopsis. The expression patterns of both enzyme families are spatiotemporally regulated during plant development [63,67–69]. Mutations affecting *TAA* and *YUC* have been suitable to demonstrate the importance of Trp-dependent IAA biosynthesis during the onset of zygotic embryogenesis [54,70,71]. As a demonstration, *YUC1,4,10,11* redundantly regulate the Arabidopsis embryonic development by modulating auxin biosynthesis at the globular stage [71]. On the other hand, the production of auxins in stamens and gynoecium has been recently reviewed [72]. During the first phases of embryo sac configuration, *TAA1*, *PIN* and *YUC* are expressed in the chalazal region of the ovule primordium, where the integuments later arise [73–75]. The reduction in *TAA1* expression results in losses of ovules, whereas the ovules are small and scarce in *pin1* mutants [73,74]. In addition, the embryo sac cellularization is directly dependent on the setting of an auxin gradient inside the sac (i.e., the highest auxin concentration originates synergids, followed by egg cell and finally, the lowest concentrations originate central cell and antipodals) [73]. Therefore, it is likely there are two key roles for auxin during ovule development: induction of embryo sac development, and control of gametophyte cell differentiation and specification [73]. Altogether, auxin seems to be fundamental for ovule development since its onset [73,75].

Given the small size of both pre- and recent-fertilized ovules, the molecular and hormonal processes that take place in them are barely known. To progress in this subject, an auxin signaling sensor named DII-VENUS was developed. Thus, high-resolution spatio-temporal information about hormone distribution and response during plant growth and development was achieved [76]. Later, in addition to the quantification of auxin through the use of antibodies, various reporter genes (e.g., *DR5v2*) have been developed to track auxin transport, level and signaling in different zygotic embryogenic tissues. Together, the auxin knowledge advanced considerably during onset fertilization [77]. As a demonstration, in the ovule of Arabidopsis and other species, DR5rev::GFP was detected in the young ovule primordia, subsequently in the tip of the nucellus and weakly in the funiculus. DR5::GFP signal is weak or undetectable before pollination or in unpollinated controls [55]. After fertilization, which induces an increase of reporter gene about 7 fold, DR5rev::GFP was localized in the integuments adjoining the micropyle and near the chalazal end of the fertilized ovule [55,78,79]. Taking together all these supporting findings [55,56,67,72], it may be pointed out that: (i) pollination leads to increased auxin levels in the fertilized maternal tissues surrounding the embryo and a localized upregulation of auxin response in the embryo attachment region. That is, auxin plays an essential role after fertilization; (ii) in early globular proembryo (8-cell embryo), YUC3,4,9 constitute the auxin biosynthetic machinery in the top suspensor cell (i.e., basal part of the embryo proper) [67], and PIN7 represents the transport machinery that delivers auxin from the suspensor to specify the proembryo (Figure 2). More specifically, PIN7 is suspensor specific and is polarized toward the proembryo, where the auxin response maximum is established [67]; (iii) *TAA1* and *YUC1, 4* genes are expressed in few apical cells of globular proembryo state (16-cell embryo), constituting another place of auxin accumulation and triggers polarization of the PIN1 proteins, but not auxin signaling. These processes contribute to the specification of the proembryo basal pole; (iv) the findings (ii) and (iii) generate an apical-basal auxin gradient provoked by the polarized localization of the auxin efflux transporters PIN1 in Arabidopsis and maize [55,80] and PIN7 in Arabidopsis [55]; (v) additionally, at a late globular stage, *YUC1* and *YUC4* were expressed in the same embryonic apical area, and *YUC8* was detected closer to the root pole. Mutations in *YUC8* lead to mitotic arrest during female gametophyte development; (vi) loss of function of *TAA1/TAR* and *YUC* genes greatly disturbs embryo development [72]. In summary, since *TAA1/TAR* and *YUC* have a tightly controlled expression, it constitutes a means of regulating the spatiotemporal auxin production within concrete tissues of the fertilized ovule. Interestingly, despite its different ovule organization, a similar increase in auxin response in fertilized maternal tissues in Arabidopsis and maize was found [55]. This fact suggests an evolutionarily conserved auxin response. On the other hand, CKs promote auxin biosynthesis genes in various organs and the appropriate ratio of auxin/CKs is important for embryo development [81]. Thus, studies shed light on how auxin and CKs interact with each other to promote and maintain the development of the gynoecium [82]. Likewise, it has been reported that auxin signaling directly activates the transcription of the CKs response regulator genes (e.g., ARABIDOPSIS RESPONSE REGULATOR7 (*ARR7*) and *ARR15*) to reduce the CKs response during early embryogenesis [83]. Besides, CKs affect the apical-basal fertilized gynoecium patterning in a similar way to the inhibition of polar auxin transport [84]. Interestingly, CKs accumulate in the proximal region of the ovule primordium in Arabidopsis [85] and exogenous CKs increase *PIN1* expression [86]. This increased *PIN1* expression is reduced in *cytokinins response factor* (*crf*) mutants [87]. Together, given the small number of results to elucidate hormonal control in the onset embryogenesis (i.e., 1-cell, 2,4-cell, octant and dermatogen), it does not seem unfortunate to venture that CKs take an important part in this regulation.

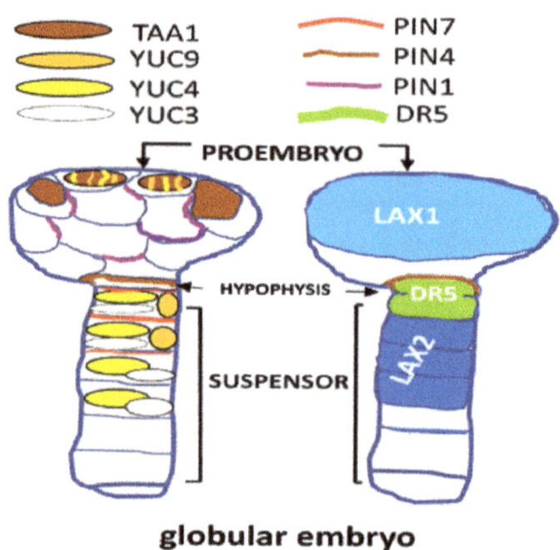

Figure 2. Dynamic of expression and localization (i.e., proembryo, hypophysis and suspensor) corresponding to several genes for the biosynthesis and transport of auxins in Arabidopsis embryos at the globular stage. LIKE-AUX1/2 (LAX1/2); highly active synthetic auxin response element (AuxRE), is referred to as DR5; auxin efflux carrier PIN-FORMED (PIN); YUCCA (YUC); TRYPTOPHAN AMINOTRANSFERASE OF ARABIDOPSIS (TAA1).

Finally, recent studies in ovules, mainly in Arabidopsis, have shown that *PIN1* expression was observed in the distal nucellus regions, showing polar localization in epidermal cells, which likely coincides with the accumulation of auxin in the ovule tip prior to megasporogenesis [79,88,89]. Besides embryo cells and suspensor, a third source of auxin can be located in funiculus [88]. Interestingly, the Trp-independent IAA biosynthetic pathway, which involves the cytosol-localized indole synthase (INS), is critical for apical-basal pattern formation during early embryogenesis in Arabidopsis [90]. Likewise, through genetic, biochemical, and functional studies, it was recently evidenced that the coordinated action of biosynthetic pathways of IAA-dependent and independent of Trp regulates the zygotic embryogenesis in Arabidopsis [46,71,90]. The auxin production via IPA is preferably involved in embryogenesis and its synthesis initially starts in the suspensor cells (i.e., the uppermost cell), and then transported into the embryo through the efflux regulator PIN7 [55,91]. Although the synthesis and localization of auxins are being widely studied, its regulation is less so. However, recent findings suggest that certain TFs may be linked to auxinic regulation during embryogenesis. Thus, the fact that one of MADS-box TFs, MADS29, a key regulator in endosperm development, is also induced by auxin in *Oryza sativa*, suggests some alterations in the auxin level during endosperm development [92].

2.2. The Hypophysis and Suspensor Identity Is Auxin-Subordinate

On the globular stage (Figure 2), when the embryogenic cells acquires its identity, the uppermost suspensor cell differentiates into the hypophysis (HP), which generates the progenitors of the quiescent center and columella stem cells, respectively [91]. That is, HP is the founder cell of the root, stem-cell system. The embryo proper plays a critical role in maintaining the identity of suspensor, which have the embryogenic potential to form a second embryo. Therefore, normally quiescent suspensor cells can develop a second embryo when the initial embryo is damaged, or when the auxin response is locally blocked. That is, through a still unknown mechanism, suspensor cells can be reprogrammed to form a second embryo. In addition to mediating HP specification, it was evidenced that auxin is also involved to maintain suspensor cell identity [93]. An auxin response maximum exists in the HP and this auxin accumulation is critical for HP differentiation. Auxin accumulation is generated by polar localization of the auxin efflux transporter PIN1 (localized to the plasma membrane) and polar auxin transport [67]. On the other hand, HP specification is transcriptionally regulated and its asymmetric division requires protein N-terminal acetylation [94]. To summarize the HP specification, MONOPTEROS (MP) activates their downstream targets, including TARGET OF MONOPTEROS 7 (TMO7), in future vascular and ground tissue cells. TMO7 moves from provascular cells to the uppermost suspensor cell. Additionally, MP promotes PIN1 to the uppermost suspensor cell. Here, both auxin responses through ARF9 and other ARFs, and TMO7 are required to specify the HP. In other words, an increase in auxin in the basal cell of the pro-embryo relieves the repression of MP and expression of TMO7, which moves into the suspensor top cell. When the HP divides, high auxin is transported to the basal daughter cell triggering the inhibition of CKs signaling through direct transcriptional activation of ARABIDOPSIS RESPONSE REGULATOR genes, *ARR7* and *ARR15* [95]. It is important to highlight that CKs are required in the apical daughter cell to specify the quiescent center. This partitioning of auxin and CKs signaling is required for the proper specification of the basal daughter cell as columella and the apical daughter cell as the quiescent center. On the other hand, the auxin response components in the pro-embryo and the suspensor are different [93]. Recently, consistent publications have already confirmed the relationship between auxin and the suspensor identity.

Thereby, (i) the embryo proper functions as an inhibitor to suppress the embryogenic potential of suspensor cells and, thus, maintains the suspensor identity during normal embryogenesis. In this process, the auxin is involved [96]. Likewise, the ribosomal gene named *RPL18aB* is responsible for maintaining suspensor cell identity in *A. thaliana* [97]. Thus, in *rpl18aB*, even when the embryo proper and suspensor were connected, the suspensor lost its identity and developed into a multicellular structure. In this process, the polar auxin transport is disturbed in the *rpl18aB* embryo [98]. Lastly, besides demonstrating the importance of auxin homeostasis in the pro-embryo-suspensor complex, Weijers' group also identified a genetic network involving several basic Helix Loop Helix (bHLH) TFs that mediate auxin action in controlling suspensor development and/or maintenance of its identity [99]. That is, bHLH TFs are involved in the suspensor auxin response. Specifically, bHLH49 appears to be a notable mediator of the auxin-dependent suppression of embryo identity in suspensor cells [99]. Interestingly, the misexpression of *bHLH49* alone induced excess divisions and even the formation of multiple embryo-like structures in suspensors, similar to the effect of inhibition of the auxin response [99]. However, the specific role of this *bHLH49* gene is under study.

2.3. Involvement of Auxins in the Coordination of Endosperm-Integuments Development

As described above, the processes of fertilization and post-fertilization of the ovule are the most studied in relation to the auxin attributions during zygotic embryogenesis. However, the knowledge of the auxin itinerary from its source to essential seed organs (i.e., integuments, endosperm and embryo) and auxin distribution in them (i.e., micropylar and chalazal regions, among others), is still far from known in detail. The use of auxin input reporters (e.g., R2D2) or auxin signaling markers (e.g., DR5; see above) is manifested by contributing to clarify both auxin itinerary and distribution. Within this complex puzzle, the more than likely TF AGAMOUS LIKE MADS-box domain (amino acids 6 to 66) protein called AGL62 plays a determinant role since it is at first involved in the development of central cell and/or endosperm [61,100]. That is, AGL62 contributes to endosperm initiation through repressing auxin biosynthesis genes expression. Data from *agl62-2* phenotype (i.e., endosperm cellularized prematurely and retains auxins) and *AGL62* expression, support the insight that AGL62 is active during the syncytial phase by suppressing the expression of genes needed for endosperm cellularization [100]. Interestingly, *agl62* seeds fail to initiate the constitution of integuments, despite the presence of dividing endosperm [100]. Accordingly, AGL62 seems to be essential for the generation of the signal that starts seed-coat development [101]; the development of endosperm and seed-coat is tightly linked to Polycomb Repressive Complex 2 (PRC2) function. Thus, removal of the endosperm inhibits seed-coat development [102]. It is to highlight that in the early stages of endosperm development, auxin appears to highly accumulate at the margin but is relatively low in the center of this organ [9]. This auxin distribution in maize endosperm is disrupted by the presence of the auxin transport inhibitor *N*-1-naphthylphthalamic acid (NPA), resulting in a multilayered aleurone [103]. At present, it is robustly established that endosperm cellularization is triggered by suppression of *AGL62* at the end of the syncytial phase and that this suppression is mediated by the Fertilization Independent Seed (FIS)-PRC2 (i.e., FIS-PCR2) [99,102,104]. Moreover, AGL62 may be regulated directly or indirectly by imprinted genes, of which genes in the FIS-PCR2 complex are the most obvious candidates. Together, after fertilization of the central cell, the endosperm initiates a signal through the action of AGL62, relieving the FIS mediated repression and leading to the differentiation of the ovule integuments into the seed-coat. In Arabidopsis, the FIS-PRC2 plays a central role in mitotic repression of the central cell and endosperm cellularization [104]. However, the pathway by which the FIS-PCR2 complex suppresses *AGL62* expression is currently unknown.

It was suggested that auxin is the putative fertilization signal that coordinates the endosperm and seed-coat development [105]. Besides, auxin has a role in the induction of endosperm proliferation in Arabidopsis showing that auxin levels are sufficient to override FIS-PRC2 suppression of secondary nucleus proliferation. However, it is still unknown how auxin intervenes in the regulation of the FIS-PRC2 complex during fertilization. On the other hand, it has been shown that the female gametophyte cellularization is directly dependent on the establishment of an auxin gradient inside gametophyte, defining thus the fates of the female gametophyte cells [73]. In addition, female gametophyte development requires both localized auxin biosynthesis and auxin import from the sporophytic ovule [106]. However, auxin alone is not sufficient to form a fully differentiated and cellularized endosperm [61,107]. Interestingly, once fertilization has been consolidated, auxin biosynthesis in the endosperm drives its own development and is responsible for starting the appearance of the seed-coat [54]. Likewise, through a poorly known mechanism, auxin is transported into the integuments to support seed-coat growth [105]. Lastly, auxin exerts its early physiological task directly in the zone where the integuments are beginning their formation [61]. Conversely, the auxin transport from seed-coat to endosperm was also found [51].

To conclude, even though the mechanism of auxin transport between sporophytic and embryonic tissues remains elusive, very recent results further consolidate the facts indicated above. Thus, the expression of *TAA1*, *PIN3* and other auxin transport proteins increases in integuments following fertilization, while the auxin signaling is very significant in both micropyle and chalazal sections of integuments [56]; to confirm, these features are prevented when auxinic transport is hindered [108]. Together, the presence of auxin within the integuments initiates its differentiation into the seed-coat. On the other hand, it was recently shown that fertilization causes restriction of auxin export through funiculus, resulting in the spread of auxin throughout the integument [108]. On the other hand, increased auxin biosynthesis in the endosperm prevents its cellularization, leading to seed arrest [19]; these results suggest that auxin determines the timing of endosperm cellularization. Finally, an epigenetic regulation signaling pathway activates auxin production in the endosperm and its transport from the endosperm to the integuments. This transport requires ABCB/PGP10 auxin efflux proteins in the endosperm, transcriptionally regulated by AGL62 [56]. Together, the data known to date suggest that auxin is specified for regulating embryo, endosperm, and seed-coat development.

3. The Auxin-Mediated Seed Dormancy and Auxin-ABA Relationship

Seed dormancy, conceptualized as the incapacity of a viable seed to complete germination despite the conditions are favorable, guarantees that the seed germinates at a suitable time [16]. Dormancy is hormonally induced, maintained and strictly regulated by the modulation of suitable hormonal signaling networks [1,11]. Abscisic acid (ABA) is the hormone known to regulate the induction and maintain PD hindering pre-harvest sprouting (i.e., viviparism) [28,31]. However, at the beginning of this century, a series of studies have committed auxin in seed maturation, PD and germination. Among other findings, earlier studies showed that application of auxin inhibited pre-harvest sprouting in wheat and promoted key tasks in equilibrating PD and seed germination rates [109]. Likewise, it was also demonstrated that ABA represses the Arabidopsis embryonic axis growth during seed germination by enhancing auxin transport and signaling and repressing the level of expression of *AXR2/IAA7* and possibly also *AXR3/IAA17* [110]. These authors, by using an auxin transport-defective mutant (*aux1-301*), show an accelerated seed to seedling transition in the presence of ABA. This experimentation interestingly points to the fact that an auxin-ABA synergistic interaction takes place in plant growth and development. That is, deficiencies in the auxin signaling pathway result in ABA modified sensitivity during seed germination. On the other hand, it is also confirmed that seed after-ripening is associated with decreased seed sensitivity to auxin [111] and that auxin signaling pathway is activated in parallel to the acquisition of seed longevity [112]. Interestingly, some genetic evidence has suggested the involvement of auxin in the maintenance of PD in Arabidopsis. For example, the reduced PD in *taa1* and *yuc1yuc6* mutants is linked to decreased ABA sensitivity [52,112]. On the other hand, the IAA level in mature seeds appeared to be linked to PD, since those mutants that have reduced IAA contents also show a reduced PD phenotype [112]. However, the mechanism by which auxin controls seed dormancy is a question not yet clearly resolved at the molecular level. Parallel, strong genetic evidence supports a model whereby ABA-mediated inhibition of seed germination requires intact auxin biosynthesis, transport, and signaling. Two notable lines of evidence follow. The first involves ABI3, a TF involved in the initiation and maintenance of the maturation phase and considered to be a major downstream component of ABA signaling. ABI3 is induced by auxin [113,114]. Though genetic and biochemical evidence has shown that ABI3 is required for auxin-activated seed dormancy [52]. Parallel, it was also demonstrated that seeds of the Arabidopsis *abi4* and *abi5* mutants are insensitive to auxin treatment during germination, indicating that ABI4 and ABI5 are important regulators of auxin-mediated inhibition of seed germination [115]. The second evidence indicates that auxin promotes PD and inhibits germination by enhancing ABA action, but together auxin and ABA act synergistically to inhibit seed germination and the auxin-mediated inhibition of seed germination is dependent on ABA [52,116]. That is, it demonstrates in Arabidopsis a molecular link through which auxin activates ABA signaling to inhibit seed germination [52]. To demonstrate the existence of this link, several facts were proven: (i) *YUC1*, *YUC2*, and *YUC6* expression peaks during the later stages of seed development; consequently, it is more than likely that the auxin biosynthesis enhances during seed maturation; recently, *yuc1yuc6* mutant displays a significantly decreased level of PD and premature germination [112]; (ii) ABA function in seed germination is largely dependent on the TIR1/AFB-AUX/IAA-ARF–mediated auxin signaling pathway; (iii) the enhancement of PD by auxin and the inhibition of seed germination by the ABA is dependent on the function of ABI3 whose transcripts are high in dormant seeds but low after germination. At the evolutionary level, it is interesting that the auxin regulatory mechanism evidenced by the He's group [52] was later found and conserved in liverworts. This last work demonstrates that endogenous auxin works as a positive regulator of liverworts *Marchantia polymorpha* gemmae dormancy. Summarizing, all these shown facts clearly reveal an unequivocal positive correlation between auxin content/signaling and PD, as it was previously demonstrated also for ABA [117].

In considering the above evidence, it is clearly concluded that the manner in which auxin acts on PD is far from known. On the other hand, it is well-founded that during seed germination, the microRNA miR160 is involved in the regulation of auxin-ABA crosstalk reducing thus the ABA effect [118].

This work demonstrates that miR160 negatively regulates ARF10 and the ABA hypersensitivity of *mARF10* (i.e., miR160-resistant form of ARF10) mutant seeds was mimicked in wild-type plants by exogenous auxin [118]. In relation to this last finding, when auxin levels are high, the auxin-responsive TFs ARF10 and ARF16 (i.e., positive regulators) which are targeted by miR160, indirectly promote the ABI3 transcription, and consequently maintain PD levels and repress germination. In other words, since ARF10 and ARF16 likely do not directly bind to the ABI3 promoter [52], they may recruit or activate an additional seed-specific TF(s) to stimulate *ABI3* expression. Conversely, at low auxin levels, ARF10 and ARF16 are repressed by AXR2/AXR3 [110]. Genetic evidence indicates that arf10 and arf1 mutants display resistance to ABA in germination assays, whereas those defective in the transcriptional suppressor ARF2 (i.e., arf2) display hypersensitivity [52,119]. That is, ARF2 is a negative regulator in ABA-mediated seed germination [119]. Lately, it was demonstrated that Germostatin Resistance Locus 1 (GSR1), encoding a tandem plant homeodomain (PHD) finger protein, forms a co-repressor with ARF16 to regulate seed germination. GSR1 physically interacts with ARF16 to possibly make up an unknown still complex functioning in auxin signaling to regulate gene transcription. This compelling finding indicates that GSR1 may be a member of an auxin-mediated seed germination genetic network [120]. In this thorough work and through chemical–genetic screenings was demonstrated that the germostatine (GS) is a small non-auxin molecule that mimics the effects of auxin and inhibits seed germination and specifically acts on auxin-mediated seed germination [120]. Given the notable characteristics of GS, its research has great prospects at both the genetic and molecular levels. Recently, a preliminary, but a striking study in barley provided evidence for miR393-mediated regulation of auxin response and its interaction with the ABA and GAs pathways during seed development and germination [121,122]. Finally, further screening of dormancy mutants is needed to identify the missing link in the ARF10/ARF16–ABI3 signaling cascade.

Overall, all the research done so far on the auxin-ABA interrelationship opens up a lot of objectives, all of which are attractive. Some of them are included in the following section. However, does auxin have any effect on ABA synthesis and GA biosynthesis/signaling pathways? If so, how does this affect take place? A comprehensive analysis of the auxin responsiveness of ABA biosynthesis, transport, and signaling mutants will be required to determine whether ABA acts downstream in any auxin-regulated process and it should be reinforced through genetic and molecular approaches.

4. Future Perspectives

The main factors involved in the induction and maintenance of PD have presumably been described. Thus, at the end of the 20th century and in these last two decades, a lot of experimentation was done to try to understand and explain the key role of ABA in the PD process. Evidence accumulated so far indicates that a concerted action of endogenous signals and environmental cues is required for PD to manifest at the end of seed development. Whereby, it is necessary to continue scrutinizing to know how endogenous and exogenous cellular signals regulate the work of the ABA. Therefore, the identification of major genetic and molecular factors is being investigated in detail during seed development, seed storage, and germination. During the study of auxin involvement in plant immunity, it was evidenced that auxin protects and strictly regulates PD through enhancing ABA signal transduction, identifying auxin as a promoter of PD [51,52]. These findings were supported, among other consistent experimentations, by the dormancy variation among seeds with altered auxin synthesis genes. Given that L-Trp-independent auxin biosynthesis contributes to the development of embryogenesis in Arabidopsis, it will be of great interest to investigate its possible participation in PD and its relation with ABA signaling. In addition, the signaling pathway linking auxin/ABI3/PD was hypothesized to be a consequence of the recruitment of ARF10/ARF16 to control the *ABI3* expression. A lot of research will be essential to identify the IAA/Aux/ARF combination that leads specifically to the activation of ABI3. After the very interesting results of Z.H. He's group, a large number of questions are emerging. Thus, do the same signals affect auxin synthesis and signaling to regulate PD? Further expanding the question, do ABA and auxin affect the homeostasis of other hormones?

Regarding this matter, it will be interesting to investigate how *YUC* genes (e.g., *YUC1, YUC2, YUC6*) are regulated to fine-tune auxin biosynthesis during seed maturation. In other words, a lot of data is lacking at the molecular level to have a coherent understanding of the interaction between ABA/auxin biosynthetic pathways. However, a fact seems shown, ABA induces the synthesis of auxin. Therefore, it will be important to analyze the importance of regulation of hormonal conjugation, degradation, and control of overlapping gene sets. This thorough and complex mechanism for IAA homeostasis is still starting to understand. In addition, one approach of great interest is, can auxin provoke cellular responses that differ according to their cellular concentrations? If so, this fact would give an added value to the determining role of auxin in plant development. On the other hand, recent information demonstrates that auxin acts downstream of ABA to promote a process (e.g., root hair elongation and seed germination). However, it is unknown if there is any process in which the ABA acts downstream of auxin. The study of the possible points of auxin-ABA interaction (e.g., does auxin affect ABI4 and ABI5?) in the regulation of different plant growth and developmental processes is still in its infancy. If so, this fact would give an added value to the determining role of auxin in plant development. On the other hand, recent information demonstrates that auxin acts downstream of ABA to promote a process (e.g., root hair elongation and seed germination). However, it is unknown if there is any process in which the ABA acts downstream of auxin. The study on the possible points of auxin-ABA interaction (e.g., does auxin affect ABI4 and ABI5?) in the regulation of different plant growth and developmental processes is still in its infancy. As is also the intervention of auxin in the seed after the ripening process.

Funding: This research received no external funding.

Acknowledgments: I sincerely acknowledge Zuhua He, National Laboratory of Plant Molecular Genetics, Institute of Plant Physiology and Ecology, Shanghai Institutes for Biological Sciences, Chinese Academy of Sciences, 300 Fenglin Road, Shanghai 200032, China, for sending manuscripts, critical reading, discussion and constructive comments of this Review. I apologize to those colleagues whose work was not included here.

Conflicts of Interest: The authors declare no conflict of interest.

References

1. Nonogaki, H. Seed germination and dormancy: The classic story, new puzzles, and evolution. *J. Int. Plant Biol.* **2019**, *61*, 541–563. [CrossRef] [PubMed]
2. Dekkers, B.J.; Pearce, S.; van Bolderen-Veldkamp, R.P.; Marshall, A.; Widera, P.; Gilbert, J.; Drost, H.G.; Bassel, G.W.; Müller, K.; King, J.R.; et al. Transcriptional dynamics of two seed compartments with opposing roles in Arabidopsis seed germination. *Plant Physiol.* **2013**, *163*, 205–215. [PubMed]
3. Steinbrecher, T.; Leubner-Metzger, G. Tissue and cellular mechanics of seeds. *Curr. Opin. Genet. Dev.* **2018**, *51*, 1–10. [CrossRef]
4. Endress, P.K. Angiosperm ovules: Diversity, development, evolution. *Ann. Bot.* **2011**, *107*, 1465–1489. [PubMed]
5. Coen, O.; Magnani, E. Seed-coat thickness in the evolution of angiosperms. *Cell. Mol. Life Sci.* **2018**, *75*, 2509–2518. [CrossRef]
6. Hehenberger, E.; Kradolfer, D.; Köhler, C. Endosperm cellularization defines an important developmental transition for embryo development. *Development* **2012**, *139*, 2031–2039. [CrossRef] [PubMed]
7. Batista, R.A.; Figueiredo, D.D.; Santos-González, J.; Khöler, C. Auxin regulates endosperm cellularization in Arabidopsis. *Genes Dev.* **2019**, *33*, 466–476.
8. Nowack, M.K.; Ungru, A.; Bjerkan, K.N.; Grini, P.E.; Schnittger, A. Reproductive cross-talk: Seed development in flowering plants. *Biochem. Soc. Trans.* **2010**, *38*, 604–612.
9. Locascio, A.; Roig-Villanova, I.; Bernardi, J.; Varotto, S. Current perspectives on the hormonal control of seed development in Arabidopsis and maize: A focus on auxin. *Front. Plant Sci.* **2014**, *5*, 412.
10. Radchuk, V.; Borisjuk, L. Physical, metabolic and developmental functions of seed-coat. *Front. Plant Sci.* **2014**, *5*, 510. [CrossRef]
11. Shu, K.; Liu, X.D.; Xie, Q.; He, Z.H. Two faces of one seed: Hormonal regulation of dormancy and germination. *Mol. Plant* **2016**, *9*, 34–45. [CrossRef]

12. Tvorogova, V.E.; Lutova, L.A. Genetic regulation of zygotic embryogenesis in angiosperm plants. *Russ. J. Plant Physiol.* **2018**, *65*, 1–14. [CrossRef]

13. Sun, X.; Shantharaj, D.; Kang, X.; Ni, M. Transcriptional and hormonal signaling control of Arabidopsis seed development. *Curr. Opin. Plant Biol.* **2010**, *13*, 611–620. [CrossRef] [PubMed]

14. Fatihi, A.; Zbierzak, A.M.; Dörmann, P. Alterations in seed development gene expression affect size and oil content of Arabidopsis seeds. *Plant Physiol.* **2013**, *163*, 973–985. [CrossRef] [PubMed]

15. Doll, N.M.; Depège-Fargeix, N.; Rogowsky, P.M.; Widiez, T. Signaling in early maize kernel development. *Mol. Plant* **2017**, *10*, 375–388. [CrossRef] [PubMed]

16. Penfield, S. Seed dormancy and Germination. *Curr. Biol.* **2017**, *27*, R874–R878. [CrossRef]

17. Chahtane, H.; Kim, W.; López-Molina, L. Primary seed dormancy: A temporally multilayered riddle waiting to be unlocked. *J. Exp. Bot.* **2017**, *68*, 857–869. [CrossRef]

18. Matilla, A.J. Programmed cell death in seeds: An adaptive mechanism required for life. In *Seed Dormancy and Germination*; IntechOpen: London, UK, 2019. [CrossRef]

19. Dong, T.; Park, Y.; Hwang, I. TAbscisic acid: Biosynthesis, inactivation, homoeostasis and signalling. *Essays Biochem.* **2015**, *58*, 29–48. [CrossRef]

20. Hauser, F.; Li, Z.; Waadt, R.; Schroeder, J.I. SnapShot: Abscisic acid signaling. *Cell* **2017**, *171*, 1708. [CrossRef]

21. Matilla, A.J.; Carrillo-Barral, N.; Rodríguez-Gacio, M.C. An update on the role of NCED and CYP707A ABA metabolism genes in seed dormancy induction and the response to after-ripening and nitrate. *J. Plant Growth Regul.* **2015**, *34*, 274–293. [CrossRef]

22. Kanno, Y.; Jikumaru, Y.; Hanada, A.; Nambara, E.; Abrams, S.R.; Kamiya, Y.; Seo, M. Comprehensive hormone profiling in developing Arabidopsis seeds: Examination of the site of ABA biosynthesis, ABA transport and hormone interactions. *Plant Cell Physiol.* **2010**, *51*, 1988–2001. [CrossRef] [PubMed]

23. Seo, M.; Marion-Poll, A. Abscisic acid metabolism and transport. In *Advances in Botanical Research*; Elsevier Inc.: Amsterdam, The Netherlands, 2019; Volume 92, pp. 1–49. ISSN 0065-2296.

24. Lefebvre, V.; North, H.; Frey, A.; Sotta, B.; Seo, M.; Okamoto, M.; Nambara, E.; Marion-Poll, A. Functional analysis of Arabidopsis NCED6 and NCED9 genes indicates that ABA synthesized in the endosperm is involved in the induction of seed dormancy. *Plant J.* **2006**, *45*, 309–319. [CrossRef]

25. Soltani, E.; Baskin, J.M.; Baskin, C.C. A review of the relationship between primary and secondary dormancy, with reference to the volunteer crop weed oilseed rape (*Brassica napus*). *Weed Res.* **2019**, *59*, 5–14.

26. Liu, X.; Hou, X. Antagonistic regulation of ABA and GA in metabolism and signaling pathways. *Front. Plant Sci.* **2017**, *9*, 251. [CrossRef]

27. Tuan, P.A.; Kumar, R.; Toora, P.K.; Ayele, B.T. Molecular mechanisms underlying abscisic acid/gibberellin balance in the control of seed dormancy and germination in cereals. *Front. Plant Sci.* **2018**, *9*, 668. [PubMed]

28. Mine, A. Interactions between abscisic acid and other hormones. In *Abscisic Acid in Plants*; Seo, M., Marion-Poll, A., Eds.; Academic Press: Cambridge, MA, USA, 2019; ISBN 9780081026205.

29. Ma, Y.; Szostkiewicz, I.; Korte, A.; Moes, D.; Yang, Y.; Christmann, A.; Grill, E. Regulators of PP2C phosphatase activity function as abscisic acid sensors. *Science* **2009**, *324*, 1064–1068. [CrossRef]

30. Papacek, M.; Christmann, A.; Grill, E. Interaction network of ABA receptors in grey poplar. *Plant J.* **2017**, *92*, 199–210. [PubMed]

31. Rodríguez-Gacio, M.C.; Matilla, M.A.; Matilla, A.J. Seed dormancy and ABA signaling. *Plant Signal. Behav.* **2009**, *4*, 1035–1049. [CrossRef] [PubMed]

32. Yoshida, T.; Mogami, J.; Yamaguchi-Shinozaki, K. Omics approaches toward defining the comprehensive abscisic acid signaling network in plants. *Plant Cell Physiol.* **2015**, *56*, 1043–1052. [CrossRef]

33. Duarte, K.E.; de Souza, W.R.; Santiago, T.R.; Sampaio, B.L.; Ribeiro, A.P.; Cotta, M.G.; da Cunha, B.A.D.B.; Marraccini, P.R.N.; Kobayashi, A.K.; Molinari, H.B.C. Identification and characterization of core abscisic acid (ABA) signaling components and their gene expression profile in response to abiotic stresses in *Setaria viridis*. *Sci. Rep.* **2019**, *9*, 4028.

34. Sánchez-Vicente, I.; Albertos, P.; Lorenzo, O. Protein shuttle between nucleus and cytoplasm: New paradigms in the ABI5-dependent ABA responses. *Mol. Plant* **2019**, *12*, 1425–1427. [PubMed]

35. Zhang, J.; Hafeez, M.T.; Lei, D.D.; Zhang, W.L. Precise control of ABA signaling through post-translational protein modification. *Plant Growth Regul.* **2019**, *88*, 99–111.

36. Kamiyama, Y.; Hirotani, M.; Ishikawa, S.; Minegishi, F.; Katagiri, S.; Takahashi, F.; Nomoto, M.; Ishikawa, K.; Kodama, Y.; Tada, Y.; et al. SNF1-related protein kinase 2 directly regulate group C Raf-like protein kinases in abscisic acid signaling. *BioRxiv* **2020**. [CrossRef]

37. Carrillo-Barral, N.; Rodríguez-Gacio, M.C.; Matilla, A.J. Delay of Germination-1 (DOG1): A key to understanding seed dormancy. *Plants* **2020**, *9*, 480. [CrossRef] [PubMed]

38. Shuai, H.W.; Meng, Y.J.; Luo, X.F.; Chen, F.; Qi, Y.; Yang, W.Y.; Shu, K. The roles of auxin in seed dormancy and germination. *Yi Chuan Hered.* **2016**, *38*, 314–322.

39. Nishimura, N.; Tsuchiya, W.; Moresco, J.J.; Hayashi, Y.; Satoh, K.; Kaiwa, N.; Irisa, T.; Kinoshita, T.; Schroeder, J.I.; Yates, J.R., III; et al. Control of seed dormancy and germination by DOG1-AHG1 PP2C phosphatase complex via binding to heme. *Nat. Commun.* **2018**, *9*, 2132.

40. Dekkers, B.J.; He, H.; Hanson, J.; Willems, L.A.; Jamar, D.C.; Cueff, G.; Rajjou, L.; Hilhorst, H.W.; Bentsink, L. The Arabidopsis *DELAY OF GERMINATION 1* gene affects *ABSCISIC ACID INSENSITIVE 5 (ABI5)* expression and genetically interacts with *ABI3* during Arabidopsis seed development. *Plant J.* **2016**, *85*, 451–465.

41. Bentsink, L.; Jowett, J.; Hanhart, C.J.; Koornneef, M. Cloning of DOG1, a quantitative trait locus controlling seed dormancy in Arabidopsis. *Proc. Natl. Acad. Sci. USA* **2006**, *103*, 17042–17047.

42. Nakabayashi, K.; Bartsch, M.; Xiang, Y.; Miatton, E.; Pellengahr, S.; Yano, R.; Seo, M.; Soppe, W.J.J. The time required for dormancy release in *Arabidopsis* is determined by DELAY OF GERMINATION 1 protein levels in freshly harvested seeds. *Plant Cell* **2012**, *24*, 2826–2838. [CrossRef]

43. Finch-Savage, W.E.; Footitt, S. Seed dormancy cycling and the regulation of dormancy mechanisms to time germination in variable field environments. *J. Exp. Bot.* **2017**, *68*, 843–856. [CrossRef]

44. Li, X.; Chen, T.; Li, Y.; Wang, Z.; Cao, H.; Chen, F.; Li, Y.; Soppe, W.J.J.; Li, W.; Liu, Y. ETR1/RDO3 regulates seed dormancy by relieving the inhibitory effect of the ERF12-TPL complex on DELAY OF GERMINATION1 expression. *Plant Cell* **2019**, *31*, 832–884. [CrossRef] [PubMed]

45. Emenecker, R.J.; Strader, L.C. Auxin-abscisic acid interactions in plant growth and development. *Biomolecules* **2020**, *10*, 281.

46. Kasahara, H. Current aspects of auxin biosynthesis in plants. *Biosci. Biotechnol. Biochem.* **2016**, *80*, 34–42. [CrossRef] [PubMed]

47. Matilla, M.A.; Daddaoua, A.; Chini, A.; Morel, B.; Krell, T. An auxin controls bacterial antibiotics production. *Nucleic Acids Res.* **2018**, *46*, 11229–11238. [CrossRef]

48. Casanova-Sáez, R.; Voß, U. Auxin metabolism controls developmental decisions in land plants. *Trends Plant Sci.* **2019**, *24*, 741–754. [CrossRef]

49. Zhao, Y. Essential roles of local auxin biosynthesis in plant development and in adaptation to environmental changes. *Annu. Rev. Plant Biol.* **2018**, *69*, 417–435. [CrossRef] [PubMed]

50. Bernardi, J.; Lanubile, A.; Li, Q.B.; Kumar, D.; Kladnik, A.; Cook, S.D.; Ross, J.J.; Marocco, A.; Chourey, P.S. Impaired auxin biosynthesis in the defective endosperm18 mutant is due to mutational loss of expression in the ZmYuc1 gene encoding endosperm-specific YUCCA1 protein in maize. *Plant Physiol.* **2012**, *160*, 1318–1328. [CrossRef]

51. Chen, J.; Lausser, A.; Dresselhaus, T. Hormonal responses during early embryogenesis in maize. *Biochem. Soc. Trans.* **2014**, *42*, 325–331. [CrossRef]

52. Liu, X.; Zhang, H.; Zhaob, Y.; Fenga, Z.; Lia, Q.; Yangc, H.Q.; Luand, S.; Lie, J.; He, Z.H. Auxin controls seed dormancy through stimulation of abscisic acid signaling by inducing ARF-mediated ABI3 activation in Arabidopsis. *Proc. Natl. Acad. Sci. USA* **2013**, *110*, 15485–15490. [CrossRef]

53. Shuai, H.; Meng, Y.; Luo, X.; Chen, F.; Zhou, W.; Dai, Y.; Qi, Y.; Du, J.; Yang, F.; Liu, J.; et al. Exogenous auxin represses soybean seed germination through decreasing the gibberellin/abscisic acid (GA/ABA) ratio. *Sci. Rep.* **2017**, *7*, 12620. [PubMed]

54. Smit, M.E.; Weijers, D. The role of auxin signaling in early embryo pattern formation. *Curr. Opin. Plant Biol.* **2015**, *28*, 99–105. [CrossRef] [PubMed]

55. Robert, H.S.; Park, C.; Gutièrrez, C.L.; Wójcikowska, B.; Pěnčík, A.; Novák, O.; Chen, J.; Grunewald, W.; Dresselhaus, T.; Friml, J.; et al. Maternal auxin supply contributes to early embryo patterning in Arabidopsis. *Nat. Plants* **2018**, *4*, 548–553. [CrossRef] [PubMed]

56. Robert, H.S. Molecular communication for coordinated seed and fruit development: What can we learn from auxin and sugars? *Int. J. Mol. Sci.* **2019**, *20*, 936. [CrossRef] [PubMed]

57. Weijers, D.; Wagner, D. Transciptional responses to the auxin hormone. *Ann. Rev. Plant Biol.* **2016**, *67*, 539–574. [CrossRef]

58. Leyser, O. Auxin signaling. *Plant Physiol.* **2018**, *176*, 465–479. [CrossRef]

59. Peris, C.I.; Rademacher, E.H.; Weijers, D. Chapter one-green beginnings—Pattern formation in the early plant embryo. *Curr. Top. Dev. Biol.* **2010**, *91*, 1–27.

60. Chen, J.; Strieder, N.; Krohn, N.G.; Cyprys, P.; Sprunck, S.; Engelmann, J.C.; Dresselhaus, T. Zygotic genome activation occurs shortly after fertilization in maize. *Plant Cell* **2017**, *29*, 2106–2125.

61. Figueiredo, D.D.; Köhler, C. Auxin: A molecular trigger of seed development. *Genes Dev.* **2018**, *32*, 479–490. [CrossRef]

62. Matthes, M.S.; Best, N.B.; Robil, J.M.; Malcomber, S.; Gallavotti, A.; McSteen, P. Auxin evodevo: Conservation and diversification of genes regulating auxin biosynthesis, transport, and signaling. *Mol. Plant* **2019**, *12*, 298–320. [CrossRef]

63. Lau, S.; Slane, D.; Herud, O.; Kong, J.; Jürgens, G. Early embryogenesis in flowering plants: Setting up the basic body pattern. *Annu. Rev. Plant Biol.* **2012**, *63*, 483–506. [CrossRef]

64. Swarup, R.; Bhosale, R. Developmental roles of AUX1/LAX auxin influx carriers in plants. *Front. Plant Sci.* **2019**, *10*, 1306. [CrossRef] [PubMed]

65. Luo, J.; Zhou, J.J.; Zhang, J.Z. Aux/IAA gene family in plants: Molecular structure, regulation and function. *Int. J. Mol. Sci.* **2018**, *19*, 259. [CrossRef] [PubMed]

66. Mateo-Bonmati, E.; Casanova-Saenz, R.; Ljung, K. Epigenetic regulation of auxin homeostasis. *Biomolecules* **2019**, *9*, 623. [CrossRef] [PubMed]

67. Friml, J.; Vieten, A.; Sauer, M.; Weijers, D.; Schwarz, H.; Hamann, T.; Offringa, R.; Jürgens, G. Efflux-dependent auxin gradient establish the apical-basal axis of Arabidopsis. *Nature* **2003**, *426*, 147–153. [CrossRef]

68. Mashiguchi, K.; Tanaka, K.; Sakai, T.; Sugawara, S.; Kawaide, H.; Natsume, M.; Hanada, A.; Yaeno, T.; Shirasu, K.; Yao, H.; et al. The main auxin biosynthesis pathway in Arabidopsis. *Proc. Natl. Acad. Sci. USA* **2011**, *108*, 18512–18517. [CrossRef]

69. Stepanova, A.N.; Yun, J.; Robles, L.M.; Novak, O.; He, W.; Guo, H.; Ljung, K.; Alonso, J.M. The Arabidopsis YUCCA1 flavin monooxygenase functions in the indole-3-pyruvic acid branch of auxin biosynthesis. *Plant Cell* **2011**, *23*, 3961–3973. [CrossRef]

70. Ljung, L. Auxin metabolism and homeostasis during plant development. *Development* **2013**, *140*, 943–950. [CrossRef]

71. Cheng, Y.; Dai, X.; Zhao, Y. Auxin synthesized by the YUCCA flavin monooxygenases is essential for embryogenesis and leaf formation in Arabidopsis. *Plant Cell* **2007**, *19*, 2430–2439. [CrossRef]

72. Robert, H.S.; Khaitova, C.L.; Mroue, S.; Benková, E. The importance of localized auxin production for morphogenesis of reproductive organs and embryos in Arabidopsis. *J. Exp. Bot.* **2015**, *66*, 5029–5042. [CrossRef]

73. Pagnussat, G.C.; Alandete-Saez, M.; Bowman, J.L.; Sundaresan, V. Auxin-dependent patterning and gamete specification in the Arabidopsis female gametophyte. *Science* **2009**, *324*, 1684–1689. [CrossRef]

74. Nole-Wilson, S.; Azhakanandam, S.; Franks, R.G. Polar auxin transport together with aintegumenta and revoluta coordinate early *Arabidopsis* gynoecium development. *Dev. Biol.* **2010**, *346*, 181–195. [CrossRef] [PubMed]

75. Bencivenga, S.; Colombo, L.; Masiero, S. Angiosperms ovules: Diversity, development, evolution. *Sex. Plant Reprod.* **2011**, *24*, 113–121. [CrossRef] [PubMed]

76. Brunoud, G.; Wells, D.M.; Oliva, M.; Larrieu, A.; Mirabet, V.; Burrow, A.H.; Beeckman, T.; Kepinski, S.; Traas, J.; Bennett, M.J.; et al. A novel sensor to map auxin response and distribution at high spatio-temporal resolution. *Nature.* **2012**, *482*, 103–106. [PubMed]

77. Liao, C.Y.; Smet, W.; Brunoud, G.; Yoshida, S.; Vernoux, T.; Weijers, D. Reporters for sensitive and quantitative measurement of auxin response. *Nat. Methods* **2015**, *12*, 207–210. [CrossRef]

78. Vanneste, S.; Friml, J. Auxin: A trigger for change in plant development. *Cell* **2009**, *136*, 1005–1016. [CrossRef]

79. Shirley, N.J.; Aubert, M.K.; Wilkinson, L.G.; Bird, D.C.; Lora, J.; Yang, X.; Tucker, M.R. Translating auxin responses into ovules, seeds and yield: Insight from Arabidopsis and the cereals. *J. Integr. Plant Biol.* **2019**, *61*, 310–336. [CrossRef]

80. Prigge, M.J.; Platre, M.; Kadakia, N.; Zhang, Y.; Greenham, K.; Szutu, W.; Pandey, B.K.; Bhosale, R.A.; Bennett, M.J.; Busch, W.; et al. Genetic analysis of the Arabidopsis TIR1/AFB auxin receptors reveals both overlapping and specialized functions. *eLife* **2020**, *9*, e54740. [CrossRef]

81. Müller, C.J.; Larsson, E.; Spíchal, L.; Sundberg, E. Cytokinin-auxin crosstalk in the gynoecial primordium ensures correct domain patterning. *Plant Physiol.* **2017**, *175*, 1144–1157. [CrossRef]

82. Reyes-Olalde, J.I.; Zúñiga-Mayo, V.M.; Serwatowska, J.; Montes, R.A.C.; Lozano-Sotomayor, P.; Herrera-Ubaldo, H.; Gonzalez-Aguilera, K.L.; Ballester, P.; Ripoll, J.J.; Ezquer, I.; et al. The bHLH transcription factor SPATULA enables cytokinin signaling, and both activate auxin biosynthesis and transport genes at the medial domain of the gynoecium. *PLoS Genet.* **2017**, *13*, e1006726. [CrossRef]

83. Müller, B.; Sheen, J. Cytokinin and auxin interaction in root stem-cell specification during early embryogenesis. *Nature* **2008**, *453*, 1094–1097. [CrossRef]

84. Zúñiga-Mayo, V.M.; Reyes-Olalde, J.; Marsch-Martinez, N.; Folter, S. Cytokinin treatments affect the apical-basal patterning of the Arabidopsis gynoecium and resemble the effects of polar auxin transport inhibition. *Front. Plant Sci.* **2014**, *5*, 191. [PubMed]

85. Bartrina, I.; Otto, E.; Strnad, M.; Werner, T.; Schmülling, T. Cytokinin regulates the activity of reproductive meristems, flower organ size, ovule formation, and, thus, seed yield in *Arabidopsis thaliana*. *Plant Cell* **2011**, *23*, 69–80. [CrossRef] [PubMed]

86. Bencivenga, S.; Simonini, S.; Benková, E.; Colombo, L. The transcription factors BEL1 and SPL are required for cytokinin and auxin signaling during ovule development in Arabidopsis. *Plant Cell* **2012**, *24*, 2886–2897. [PubMed]

87. Cucinotta, M.; Manrique, S.; Guazzotti, A.; Quadrelli, N.E.; Mendes, M.A.; Benkova, E.; Colombo, L. Cytokinin response factors integrate auxin and cytokinin pathways for female reproductive organ development. *Development* **2016**, *143*, 4419–4424. [CrossRef]

88. Bencivenga, S.; Roig-Villanova, I.; Ditengou, F.A.; Palme, K.; Simon, R.; Colombo, L.L. Maternal control of PIN1 is required for female gametophyte development in *Arabidopsis*. *PLoS ONE* **2013**, *8*, e66148.

89. Lora, J.; Herrero, M.; Tucker, M.R.; Hormaza, J.I. The transition from somatic to germline identity shows conserved and specialized features during angiosperm evolution. *New Phytol.* **2017**, *216*, 495–509.

90. Wang, B.; Chu, J.; Yu, T.; Xu, Q.; Sun, X.; Yuan, J.; Xiong, G.; Wang, G.; Wang, Y.; Li, J. Tryptophan-independent auxin biosynthesis contributes to early embryogenesis in Arabidopsis. *Proc. Natl. Acad. Sci. USA* **2015**, *112*, 4821–4826. [CrossRef]

91. Palovaara, J.; de Zeeuw, T.; Weijers, D. Tissue and organ initiation in the plant embryos: A first time for everything. *Annu. Rev. Cell Dev. Biol.* **2016**, *32*, 47–75. [CrossRef]

92. Yin, L.L.; Xue, H.W. The MADS29 transcription factor regulates the degradation of the nucellus and the nucellar projection during rice seed development. *Plant Cell* **2012**, *24*, 1049–1065. [PubMed]

93. Rademacher, E.H.; Lokerse, A.S.; Schlereth, A.; Peris, C.I.; Bayer, M.; Kientz, M.; Freire Rios, A.; Borst, J.W.; Lukowitz, W.; Jürgens, G.; et al. Different auxin response machineries control distinct cell fates in the early plant embryo. *Dev. Cell* **2012**, *22*, 211–222. [CrossRef]

94. Feng, J.; Li, R.; Yu, J.; Ma, S.; Wu, C.; Li, Y.; Cao, Y.; Ma, L. Protein N-terminal acetylation is required for embryogenesis in Arabidopsis. *J. Exp. Bot.* **2016**, *67*, 4779–4789. [CrossRef] [PubMed]

95. Schaller, G.E.; Bishopp, A.; Kieber, J.J. The Yin-Yang of Hormones: Cytokinin and Auxin Interactions in Plant Development. *Plant Cell* **2015**, *27*, 44–63. [CrossRef] [PubMed]

96. Liu, Y.; Li, X.; Zhao, J.; Tang, X.; Tian, S.; Chen, J.; Shi, C.; Wang, W.; Zhang, L.; Feng, X. Direct evidence that suspensor cells have embryogenic potential that is suppressed by the embryo proper during normal embryogenesis. *Proc. Natl. Acad. Sci. USA* **2015**, *112*, 12432–12437. [CrossRef] [PubMed]

97. Yan, H.; Chen, D.; Wang, Y.; Sun, Y.; Zhao, J.; Sun, M.; Peng, X. Ribosomal protein L18aB is required for both malegametophyte function and embryo development inArabidopsis. *Sci. Rep.* **2016**, *6*, 31195. [CrossRef] [PubMed]

98. Xie, F.; Yan, H.; Sun, Y.; Wang, Y.; Chen, H.; Mao, W.; Zhang, L.; Sun, M.; Peng, X. RPL18aB helps maintain suspensor identity duringearly embryogenesis. *J. Integr. Plant Biol.* **2018**, *60*, 266–269. [CrossRef]

99. Radoeva, T.; Lokerse, A.S.; Peris, C.I.; Wendrich, J.R.; Xiang, D.; Liao, C.-L.; Vlaar, L.; Boekschoten, M.; Hooiveld, G.; Datla, R.; et al. A robust auxin response network controls embryo and suspensor development through a basic helix loop helix transcriptional module. *Plant Cell* **2019**, *31*, 52–67. [CrossRef]

100. Kang, I.-H.; Steffen, J.G.; Portereiko, M.F.; Lloyd, A.; Drews, G.N. The AGL62 MADS domain protein regulates cellularization during endosperm development in Arabidopsis. *Plant Cell* **2008**, *20*, 635–647. [CrossRef]

101. Coen, O.; Fiume, E.; Xu, W.; De Vos, D.; Lu, J.; Pechoux, C.; Lepiniec, L.; Magnani, E. Developmental patterning of the sub-epidermal integument cell layer in Arabidopsis seeds. *Development* **2017**, *144*, 1490–1497. [CrossRef]

102. Figueiredo, D.D.; Köhler, C. Signalling events regulating seed coat development. *Biochem. Soc. Transl.* **2014**, *42*, 358–363. [CrossRef]

103. Roszak, P.; Köhler, C. Polycomb group proteins are required to couple SC initiation to fertilization. *Proc. Natl. Acad. Sci. USA* **2011**, *108*, 20826–20831. [CrossRef]

104. Forestan, C.; Meda, S.; Varotto, S. ZmPIN1-mediated auxin transport is related to cellular differentiation during *maize* embryogenesis and endosperm development. *Plant Physiol.* **2010**, *152*, 1373–1390. [CrossRef] [PubMed]

105. Hands, P.; Rabiger, D.S.; Koltunow, A. Mechanisms of endosperm initiation. *Plant Reprod.* **2016**, *29*, 215–225. [CrossRef]

106. Figueiredo, D.D.; Batista, R.A.; Roszak, P.J.; Henning, L.; Köhler, C. Auxin production in the endosperm drives seed-coat development in Arabidopsis. *eLife* **2016**, *5*, e20542. [CrossRef]

107. Panoli, A.; Martin, M.V.; Alandete-Saez, M.; Simon, M.; Neff, C.; Swarup, R.; Bellido, A.; Yuan, L.; Pagnussat, G.C.; Sundaresan, V. Auxin import and local auxin biosynthesis are required for mitotic divisions, cell expansion and cell specification during female gametophyte development in *Arabidopsis thaliana*. *PLoS ONE* **2015**, *10*, e0126164. [CrossRef] [PubMed]

108. Figueiredo, D.D.; Batista, R.A.; Roszak, P.J.; Köhler, C. Auxin production couples endosperm development to fertilization. *Nat. Plants* **2015**, *1*, 15184. [CrossRef] [PubMed]

109. Larsson, E.; Vivian-Smith, A.; Offringa, R.; Sundberg, E. Auxin homeostasis in Arabidopsis ovules is anther-dependent at maturation and changes dynamically upon fertilization. *Front. Plant Sci.* **2017**, *8*, 1735. [CrossRef]

110. Ramaih, S.; Guedira, M.; Paulsen, G.M. Relationship of indoleacetic acid and tryptophan to dormancy and preharvest sprouting of wheat. *Funct. Plant Biol.* **2003**, *30*, 939–945. [CrossRef]

111. Belin, C.; Megies, C.; Hauserova, E.; Lopez-Molina, L. Abscisic acid represses growth of the Arabidopsis embryonic axis after germination by enhancing auxin signaling. *Plant Cell* **2009**, *21*, 2253–2268. [CrossRef]

112. Liu, A.; Gao, F.; Kanno, Y.; Jordan, M.C.; Kamiya, Y.; Seo, M.; Ayele, B.T. Regulation of wheat seed dormancy by after-ripening is mediated by specific transcriptional switches that induce changes in seed hormone metabolism and signaling. *PLoS ONE* **2013**, *8*, e56570. [CrossRef]

113. Pellizzaro, A.; Neveu, M.; Lalanne, L.; Vu, B.L.; Kanno, Y.; Seo, M.; Leprince, O.; Buitink, J. A role for auxin signaling in the acquisition of longevity during seed maturation. *New Phytol.* **2020**, *225*, 284–296. [CrossRef]

114. Brady, S.M.; Sarkar, S.F.; Bonetta, D.; McCourt, P. The *Abscisic acid insensitive 3* (ABI3) gene is modulated by farnesylation and is involved in auxin signaling and lateral root development in *Arabidopsis*. *Plant J.* **2003**, *34*, 67–75. [PubMed]

115. Rohde, A.; Kurup, S.; Holdsworth, M. ABI3 emerges from the seed. *Trends Plant Sci.* **2000**, *5*, 418–419. [PubMed]

116. Skubacz, A.; Daszkowska-Golec, A.; Szarejko, I. The role and regulation of ABI5 (ABA-insensitive 5) in plant development, abiotic stress responses and phytohormone crosstalk. *Front. Plant Sci.* **2016**, *7*, 1884. [PubMed]

117. Eklund, D.M.; Ishizaki, K.; Flores-Sandoval, E.; Kikuchi, S.; Takebayashi, Y.; Tsukamoto, S.; Hirakawa, Y.; Nonomura, M.; Kato, H.; Kouno, M.; et al. Auxin produced by the indole-3-pyruvic acid pathway regulates development and gemmae dormancy in the liverwort *Marchantia polymorpha*. *Plant Cell* **2015**, *27*, 1650–1669. [CrossRef] [PubMed]

118. Nambara, E.; Okamoto, M.; Tatematsu, K.; Yano, R.; Seo, M.; Kamiya, Y. Abscisic acid and the control of seed dormancy and germination. *Seed Sci. Res.* **2010**, *20*, 55–67. [CrossRef]

119. Liu, P.P.; Montgomery, T.A.; Fahlgren, N.; Kasschau, K.D.; Nonogaki, H.; Carrington, J.C. Repression of AUXIN RESPONSE FACTOR10 by microRNA160 is critical for seed germination and post-germination stages. *Plant J.* **2007**, *52*, 133–146. [CrossRef]

120. Wang, L.; Hua, D.; He, J.; Duan, Y.; Chen, Z.; Hong, X.; Gong, Z. Auxin response factor2 (ARF2) and its regulated homeodomain gene HB33 mediated abscisic acid response in Arabidopsis. *PLoS Genet.* **2011**, *7*, e1002172. [CrossRef]

121. Ye, Y.; Gong, Z.; Lu, X.; Miao, D.; Shi, J.; Lu, J.; Zhao, Y. Germostatin resistance locus 1 encodes a PHD finger protein involved in auxin-mediated seed dormancy and germination. *Plant J.* **2016**, *85*, 3–15. [CrossRef]
122. Bai, B.; Shi, B.; Hou, N.; Cao, Y.; Meng, Y.; Bian, H.; Zhu, M.; Han, N. microRNAs participate in gene expression regulation and phytohormone cross-talk in barley embryo during seed development and germination. *BMC Plant Biol.* **2017**, *17*, 150. [CrossRef]

Review

Reactive Oxygen Species (ROS) and Nucleic Acid Modifications during Seed Dormancy

Kai Katsuya-Gaviria [1,2], **Elena Caro** [1,2], **Néstor Carrillo-Barral** [3] and
Raquel Iglesias-Fernández [1,2,*]

1   Centro de Biotecnología y Genómica de Plantas-Severo Ochoa (CBGP, UPM-INIA), Universidad Politécnica de Madrid (UPM)—Instituto Nacional de Investigación y Tecnología Agraria y Alimentaria (INIA), 28223 Pozuelo de Alarcón, Spain; kai.kgaviria@alumnos.upm.es (K.K.-G.); elena.caro@upm.es (E.C.)

2   Departamento de Biotecnología-Biología Vegetal, Escuela Técnica Superior de Ingeniería Agronómica, Alimentaria y de Biosistemas, UPM, 28040 Madrid, Spain

3   Departamento de Fisiología Vegetal, Facultad de Ciencias, Universidad da Coruña (UdC), 15008 A Coruña, Spain; n.carrillo@udc.es

*   Correspondence: raquel.iglesias@upm.es

Received: 24 April 2020; Accepted: 26 May 2020; Published: 27 May 2020

Abstract: The seed is the propagule of higher plants and allows its dissemination and the survival of the species. Seed dormancy prevents premature germination under favourable conditions. Dormant seeds are only able to germinate in a narrow range of conditions. During after-ripening (AR), a mechanism of dormancy release, seeds gradually lose dormancy through a period of dry storage. This review is mainly focused on how chemical modifications of mRNA and genomic DNA, such as oxidation and methylation, affect gene expression during late stages of seed development, especially during dormancy. The oxidation of specific nucleotides produced by reactive oxygen species (ROS) alters the stability of the seed stored mRNAs, being finally degraded or translated into non-functional proteins. DNA methylation is a well-known epigenetic mechanism of controlling gene expression. In *Arabidopsis thaliana*, while there is a global increase in CHH-context methylation through embryogenesis, global DNA methylation levels remain stable during seed dormancy, decreasing when germination occurs. The biological significance of nucleic acid oxidation and methylation upon seed development is discussed.

Keywords: after-ripening; DNA methylation; oxidation; RNA stability; seed dormancy; seed vigour; ROS

1. Introduction: Seed Dormancy

Seeds are the first world crop, and the basic knowledge applied to enhance complex traits such are dormancy, viability, and vigour is essential for food security and crop production. The seed represents the sexual reproductive entity of higher plants and its development is divided into embryogenesis, maturation, and germination. In the final stage of maturation, seeds begin to dehydrate and acquire tolerance to desiccation and eventually can enter into a latency state named primary dormancy [1]. Dormancy is defined as the incapacity of a mature, dry, and viable seed to germinate under favourable conditions. Non-dormant seeds are able to temporarily block their germination if exposed to unfavourable germination conditions upon seed imbibition. This phenomenon is referred to as secondary dormancy [2–4]. Pre-harvest sprouting (PHS) occurs when seeds lose dormancy and germinate in the mother plant before the dispersion takes place. PHS is an important problem in cereal production because it reduces crop yield and quality [5]. The time course of seed dormancy comprises induction, maintenance, and release, and it is well-known that the ABA/GA ratio governs these transitions, although other hormones such as auxins are also involved. In the last decades, several genes have been related to the seed dormancy process, such as those involved in ABA and

GA signalling and metabolism, and several transcription factors belonging to B3, bZIP, and RING finger families (ABI3, ABI5, FUS3, DESPIERTO) [6–10]. The *Delay Of Germination-1* (*DOG1*) gene, a quantitative trait locus controlling seed dormancy, is essential to stablish this process, but its specific molecular function is still unknown [11]. The depth of dormancy is directly related to DOG1 protein levels, and mutations or decreased expression of DOG1 during maturation generate seeds with reduced dormancy [12]. It has been recently reported that the DOG1 protein forms a complex with a type of phosphatase 2C named ABA HYPERSENSITIVE GERMINATION1 (AHG1), and this complex is critical for seed dormancy maintenance [13–17]. The state of the art regarding DOG1 is discussed in detail in a review included in this issue [18].

Dormancy is usually classified into the physiological, physical, chemical, morphological, and morpho- physiological categories [19]. Physiological dormancy is abundant in the soil seed bank, and its release is mainly governed by the after-ripening (AR) and stratification processes. Stratification occurs when seeds undergo a time-term of cold moisture that breaks dormancy, and after-ripening takes place when a seed gradually loses dormancy through a period of dry storage. Thus, AR determines the dormancy degree and consequently affects the crop performance in the field [20–22]. The period of dry storage is influenced by the environmental conditions experienced by the mother plant during seed development [23]. The AR period is also specific for each species and even for distinct accessions, such as occurs in the *Arabidopsis thaliana* Columbia (Col) and Cape Verde Islands (Cvi) ecotypes, where it ranges from one to six months, respectively [24]. In the dry state, the low water content in the seed prevents metabolic activity, raising the question of how dormancy release occurs at the molecular level. AR and no-AR imbibed seeds show specific gene expression patterns in *Arabidopsis thaliana* (hereafter *A. thaliana*) and in our model of study *Sisymbrium officinale* [20,25–29]. Interestingly, the seed redox state fluctuates during the AR period and the reactive oxygen species (ROS) quantity increases concomitantly with seed development, promoting dormancy release [30,31]. During the *A. thaliana* and *Hordeum vulgare* seed development, mRNAs are progressively accumulated and some of them remain in the dry seed state, where they are specifically translated depending on the environmental conditions [32–34]. The molecular mechanisms controlling the stored mRNA longevity and their specific translation during seed imbibition are widely unknown. However, chemical modifications of mRNA, such as the oxidation caused by ROS, have been related to its stability [35]. Other nucleic acid modifications, such as DNA and RNA methylation, have been also described to be critical for *A. thaliana* embryogenesis and seed viability, facts that demonstrate the importance of these chemical marks in seed biology [33,34,36]. These new findings and their relevance to seed biology make it timely to update this topic. Here, we discuss how mRNA and genomic (gDNA) oxidation and methylation control the expression levels of genes/proteins involved in seed dormancy release, with emphasis in the AR process.

2. Reactive Oxygen Species (ROS) Promote Dormancy Release and Modify Stability of Nucleic Acids

2.1. ROS Affect Dormancy Release

Reactive oxygen species (ROS), such as hydrogen peroxide (H_2O_2), superoxide ion ($O_2^{\bullet}$), and hydroxyl radicals ($OH^{\bullet}$), are normally produced during cell metabolism. ROS accumulate at different stages of seed development and have been correlated with a low degree of dormancy [31,37,38]. When ROS content reaches a threshold, seed ageing takes place. At this point an intensive degradation of nucleic acids, proteins, and phospholipids present at the cell membrane occurs [39]. Likewise, once the ROS threshold has been reached, dormancy is alleviated and the subsequent germination can be initiated [31,40]. The quantity of ROS depends on the balance between the production and scavenging carried out by the antioxidant systems. In dry seeds, the quantity of ROS increases through storage, and this production mainly occurs by non-enzymatic processes, such as lipid peroxidation and the Amadori–Maillard reaction, but also oxidoreductase enzymes such as NADH oxidases could participate [41,42]. In *A. thaliana* mature seeds, a cuticular film in the outer side of the endosperm (next to the seed coat) has been identified, and this structure could limit the permeability of the seed to outer compounds. Interestingly, when cutin biosynthesis is affected, mutant seeds display a low degree of dormancy and a high level of lipid oxidation [43]. The storage of seeds at dry conditions (AR) and elevated partial pressure of oxygen (EPPO) reduce dormancy faster than AR simply does [38]. Moreover, AR and dormancy release due to EPPO share QTLs such as *DOG1*. Since the seed levels of DOG1 are not reduced during the AR or the imbibition, post-translational changes have been proposed to regulate its action [44]. Recently, it has been reported that DOG1 binds to a heme-group. It is argued that the heme-binding site of DOG1 could act as a sensor of stimuli derived from the presence of oxygen, as it is observed in other heme-binding proteins [11,16].

ROS accumulate in seeds upon AR and dormancy release and promote germination in several species, as illustrated by studies in *A. thaliana*, *Pisum sativum*, *Helianthus annuus*, *Hordeum vulgare*, and *Triticum aestivum* [40,45–48]. In *A. thaliana* and *H. annuus* AR seeds, while ROS are localized close to the radicle tip of the embryonic axis during imbibition, they do not have a particular distribution in dormant embryos [49–51]. It is widely known that ROS act as signalling molecules, but H_2O_2 seems to be the main responsible for redox signalling, probably because of its stability. However, $OH^\bullet$ and $O_2^\bullet$ radicals are also produced and participate in radicle emergence and embryo growth probably by their participation in cell wall modification [52–54]. In AR seeds, ROS have been described to be involved in ethylene signalling and to alter the ABA/GA ratio by promoting the expression of *CYP707A* genes involved in ABA degradation and by increasing GA biosynthesis [45,55,56]. This is in agreement with our previous results in *Sisymbrium officinale*, where nitrate modifies the expression of *ABA* and *GA* metabolic genes in AR seeds [28]. In wheat and barley, while AR dry seeds accumulate H_2O_2 and singlet oxygen ($^1O_2$), non-enzymatic antioxidant content diminishes (i.e., ascorbate and glutathione [57,58]). Several enzymes that participate in ROS homeostasis have been associated with the germination and AR process, such as NADPH oxidases (respiratory burst oxidase homologues: rboh), Class III peroxidases (CIII Prx) and thioredoxins (Trxo1). AtRbohB is a major producer of $O_2^\bullet$ in germinating *A. thaliana* seeds, and the *AtrbohB* mutants fail to after-ripen and show reduced protein oxidation. Moreover, the inhibition of $O_2^\bullet$ production leads to a delay in *A. thaliana* and *Lepidium sativum* seed germination [41,59,60]. In *Oryza sativa* seeds, the *PHS9* gene encoding a CC-type glutaredoxin has been involved in the regulation of preharvest sprouting (PHS) through the integration of ROS and ABA signalling [61].

2.2. Oxidation Modifies Nucleic Acid Stability during Seed Dormancy

During the AR period, the specific oxidation of mRNAs and proteins (carbonylation) has been described in *Helianthus annuus* seeds [30]. The stored mRNAs in the dry seed can undergo mainly two fates: (i) translation to their corresponding proteins during early seed imbibition and (ii) degradation. The specific mRNA turnover is a prerequisite for dormancy breakage [15]. ROS oxidize nucleic acids at different positions affecting their stability. The 8-hydroxyguanosine (8-OHG) is the most frequent oxidative nucleoside in RNA molecules [62]. The oxidized mRNAs could undergo a ribosome-based quality control to be degraded by the no-go decay pathway (NGD) or translated into non-functional proteins [63]. The oxidation of mRNAs causes premature termination of translation or the appearance of errors in the translation process resulting in the subsequent degradation of the proteins [62]. It has been reported that the RNA polymerase can control the incorporation of 8-hydroxyguanosine (8-OHG) into the growing mRNA strand during transcription and this could be a process that regulates mRNA stability [64]. Dry seeds contain long-lived mRNAs, and the oxidation of these mRNAs has been linked to dormancy release and seed ageing [31,35]. In *Helianthus annuus* and *Triticum spp.* AR-seeds, the oxidation of stored mRNAs encoding enzymes involved in nutrient storage (α-amylase/trypsin inhibitors) has been related to dormancy release [30,65,66]). A decrease in the abundance of stored mRNAs associated to dormancy has been observed in *A. thaliana* AR-seeds, such as that encoding the DELLA protein GAI [67].

During maturation, seeds accumulate mRNAs, and some of them are degraded through the AR period. This degradation is caused presumably by oxidation, thus altering their stability. However, some mRNAs are stable and endure until germination. Those mRNAs that are translated during early seed germination are highly specific, and they seem to be selected by environmental conditions [32,35]. The question of how stored long-lived mRNAs remain stable until germination and how they are selected is poorly understood. It has been proposed that RNA binding proteins (RBPs) recruit mRNAs, protecting them during the storage period. These RBPs recognize different types of motifs in the RNA sequence, such as the DEAD-box helicase domain that can activate or repress translation [68]. The RBPs can also bind to the ribosome, thus controlling translation. Moreover, ribosomes can be stored in quiescent states, such as occurs in bacteria, yeast and animals [69]. It has been described that certain characteristics of mRNAs determine its association to ribosomes, such as transcript length, GC content, or the presence of upstream open reading frames (uORFs). It has been argued that the mRNA selection by RBPs lies in the presence of specific motifs in their sequences [70]. In *A. thaliana* dry seeds, stored mRNAs are mainly associated with monosomes that are transcribed during seed maturation and translated upon early germination [22]. Recently, a consensus motif (5′-GAAGAAGAA-3′) that is significantly overrepresented in 5′-UTR monosome-associated transcripts has been identified, suggesting that it could play a role in the recruitment of specific RNA-binding proteins [70] (Figure 1).

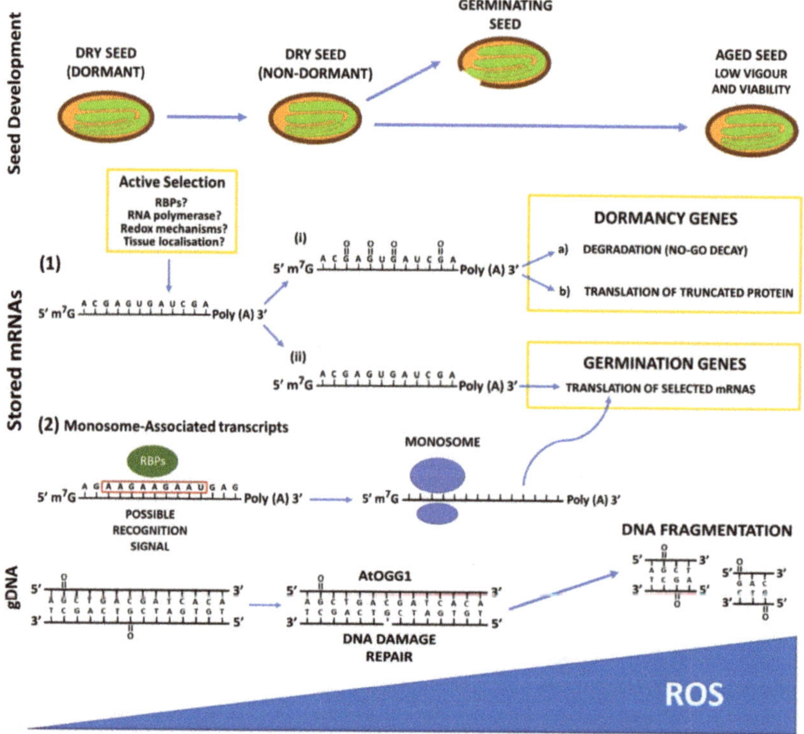

Figure 1. Schematic representation of how reactive oxygen species (ROS) affect seed dormancy and viability by the oxidation of nucleic acids (mRNA and gDNA). Blue circles depict the ribosome.

Stored seeds also suffer DNA damage caused by oxidation, which affects seed longevity. The oxidation of the guanine at the C-8 position to produce 7,8-dihydro-8-oxoguanine (8-oxoG) is the most common modification involved in DNA damage [71]. Overexpression of a DNA glycosylase/apurinic/apyrimidinic lyase (AtOGG1), involved in repairing DNA by eliminating 8-oxoG, enhances longevity in *A. thaliana* seeds [72]. ROS also affect DNA methylation by the oxidation of the methyl-5′-cytosine (m^{5}C), which is more vulnerable to oxidation [73]. During *Pyrus communis* and *Quercus robur* seed storage, DNA methylation pattern is modified, and this is correlated to the ROS levels in the seed [74,75].

3. mRNA and DNA Methylation upon Seed Development

3.1. The Seed Epitranscriptome

Non-classical chemical changes, apart from capping and polyadenylation modifications, alter the stability of mRNAs,. More than 160 chemical modifications of mRNAs have been recently reported in *A. thaliana*, as a whole they are considered the epitranscriptome [76]. These alterations include the methylation of the adenosine and cytosine nucleotides: N6-methyladenosine (m^{6}A) and 5-methylcytosine (m^{5}C) [77]. These mRNA modifications modulate transcript stability, and they have been found in transcripts that are being degraded, suggesting they are involved in mRNA turnover. However, stable mRNAs contain chemical modifications related to the alternative splicing of introns. This fact indicates that these modifications could also regulate the expression of different transcript variants [78]. mRNA chemical modification has been recently described as essential for plant development, and it is considered a new emerging level of controlling gene expression [76]. In *A. thaliana*, the first mRNA methylation map has revealed the presence of thousands of methylated mRNAs [79]. A new molecular system formed by RBPs has been involved in adding (*writers*), removing (*erasers*) and recognizing (*readers*) chemical modifications present at the mRNA transcripts. The *A. thaliana* RNA Adenosine Methylase MTA (*writer*) has been described to be essential for plant survival as null *mta* mutants are embryo-lethal [80]. MTA is mainly expressed in dividing tissues, such as seeds upon development. Moreover, the expression of MTA under the control of the seed-specific ABI3 promoter rescues the lethality of null *mta* mutants, indicating its relevance during seed development [81].

3.2. Regulation of Seed Development through DNA Methylation

3.2.1. DNA Methylation Basis

The chemical modification of DNA by the addition of a methyl group at the fifth position on the pyrimidine ring of cytosines is a stable and reversible epigenetic mark [82]. Centromeres and repeated sequences are heterochromatic regions, characterized by heavily methylated DNA [83]. DNA methylation in euchromatic regions and within genes has different effects on gene expression. While methylation within transcribed regions (the body of genes, mostly methyl-CG) is associated with constitutive expression levels [84], DNA methylation within promoters is associated with repression, i.e., gene silencing [85] and tissue-specific expression [83]. DNA methylation is a conserved epigenetic mechanism that, in plants, occurs at three different sequence contexts: the symmetrical CG and CHG, and the asymmetrical CHH, where H can be a cytosine, a thymidine or an adenine. Symmetrical methylation can be maintained during the cell cycle by DNA METHYLTRANSFERASE 1 (MET1) and CHROMOMETHYLASE 3 (CMT3), which catalyse DNA methylation on hemi-methylated DNA strands [86]. Asymmetrical methylation requires re-establishment after each DNA replication round by persistent de novo methylation by DOMAINS REARRANGED METHYLTRANSFERASE 2 (DRM2) at transposons and other repeated sequences in euchromatic chromosome arms. In this case, small RNAs cause transcriptional gene silencing by directing the addition of DNA methylation to specific DNA sequences by nucleotide homology mediated by ARGONAUTE (AGO) proteins, in a pathway called RNA-directed DNA methylation (RdDM) [87]. CHROMOMETHYLASE 2 (CMT2) catalyses CHH methylation at histone H1-containing heterochromatin, where DRM2 is inhibited [88]. Alternatively, sequences that have been methylated can become demethylated either passively or actively. Passive demethylation is due to subsequent rounds of DNA replication without the maintenance of the methylation pattern. Active demethylation is catalysed by DNA glycosylases

—DEMETER (DME), REPRESSOR OF SILENCING1/DEMETER-LIKE 1 (ROS1), DEMETER-LIKE 2 (DML2) and DEMETER-LIKE 3 (DML3), in *A. thaliana* [89].

3.2.2. DNA-Methylation and Seed Development

The initial observations of cell nuclei, during *A. thaliana* seed maturation, have shown small size and very condensed chromatin, while the contrary happens during seed germination, suggesting a change in the balance between heterochromatin and euchromatin in the transition to germination [90]. Recent whole-genome bisulfite sequencing analysis supports these previous observations and shows that DNA methylation is dynamic during the process of seed development [91–95] (Figure 2). DNA methylation is known to play a critical role in seed development since mutations in MET1 and DME have been described to affect this developmental process [96,97]. Now, we know that during seed development there is a specific increase in global non-CG methylation, mainly due to changes in methyl-CHH and especially at transposable elements (TEs), both in *Glycine max* and *A. thaliana* [91–95]. This increase in global non-CG methylation occurs simultaneously throughout the seed, while CG and CHG methylations do not change significantly during seed development or in any specific part of it [91,92,94]. In mature embryos, CHH hypermethylation in TEs matches DNA hypomethylation in the endosperm. A role for RdDM in the establishment of this methylation is supported by the fact that 24-nt siRNAs derived from these regions are most abundant during late embryogenesis, coinciding with elevated expression of RdDM components [93]. Analysis of *drm1 drm2 cmt3* (*ddc*) and *drm1 drm2 cmt3 cmt2* (*ddcc*) mutants shows that CMT2 is also involved in this TE hypermethylation by the identification of over 100 de-repressed TEs which show different CHH methylation compared to that of randomly selected TEs [91]. This fact reinforces the idea that non-CG methylation and, specifically, methyl-CHH, may be an important mechanism to avoid TE activity during seed maturation [94]. Regarding DNA methylation within genes, it has been found that those genes that are transcribed are more likely to be differentially methylated during seed maturation than non-expressed genes [92]. The majority of these differentially methylated genes become transcriptionally down-regulated as seed maturation proceeds—DNA methylation levels increase—pointing to a specific role of DNA methylation in the repression of specific genes [92].

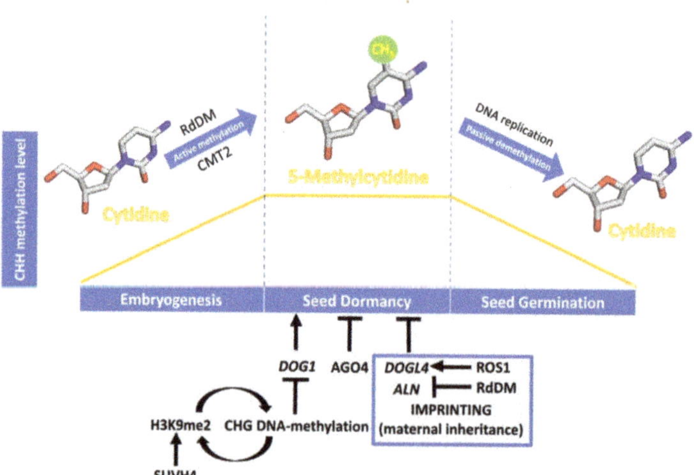

Figure 2. Dynamics of CHH DNA methylation (orange line) during seed development, and epigenetic regulation of seed dormancy. The DNA methylation process has been schematised with cytidine and 5-methylcytidine molecules designed using Chimera 1.14 [98]. Epigenetic regulation of seed dormancy has been schematised with arrows indicating a promotive effect and bars indicating a repressive effect. Regulation through genomic imprinting is indicated within the blue box.

The expression levels of *DOG1* are regulated by various molecular mechanisms like alternative splicing, selective polyadenylation, and non-coding RNAs [18,99], but in this review we will highlight the consequences of *DOG1* H3K9 and DNA methylation in the regulation of seed dormancy. DNA methylation in the CHG context is perpetuated in *A. thaliana* by the SUVH family of histone H3 lysine 9 N-methyltransferases. Their SRA domain binds to methylated CHG cytosines, leading to H3K9me2 deposition. Conversely, sites of H3K9 methylation recruit DNA methyltransferases CMT3 and CMT2, thus forming a self-reinforcing loop of repressive epigenetic marks. The seeds of SUVH4 mutants (also known as *kyp*) show enhanced dormancy and up-regulation of *DOG1* and *ABI3* [100], supporting that histone H3K9me2 caused by KYP/SUVH4 induces their silencing through DNA methylation, negatively affecting seed dormancy. A possible role of the RdDM pathway in the silencing of seed dormancy genes has been suggested from studies in cereals. In barley, the *AGO4_9* gene *AGO1003* is expressed differentially in the embryos of dormant and non-dormant grains [101]. In wheat, *AGO802B*, a wheat ortholog of an *AGO4_9*, is expressed during grain development and its expression is significantly lower in preharvest sprouting-resistant varieties than in susceptible ones; implying that DNA methylation levels could be reduced in PHS susceptible varieties compared to resistant lines [102]. This result indicates that AGO4 is a negative regulator of dormancy, although it is not known whether specific coding genes are subjected to silencing through RdDM or which ones are. Recently, it has been demonstrated that mutants with loss or reduced function of the chromatin-remodelling factor PICKLE (PKL) show increased seed dormancy; PKL directly represses *DOG1* expression [103].

Genomic imprinting is an epigenetic mechanism that alters gene expression in a parent-of-origin manner. As a consequence, the expression of some genes depends on whether the allele was inherited from the maternal or the paternal parent [104]. Reciprocal crosses between *A. thaliana* accessions with increased seed dormancy—Cvi—and with reduced seed dormancy—C24—have suggested a maternal inheritance of seed dormancy levels. Thus, the two populations of F-1 hybrid seeds from these reciprocal crosses exhibit distinct levels of dormancy, which tend to phenocopy the maternal traits. In parallel, 57 dormancy-specific maternally expressed genes have been analysed [105], but the molecular mechanisms underpinning the regulation of these genes are yet to be elucidated. Recent studies have shown the importance of genomic imprinting, through DNA-methylation, in regulating the expression in the endosperm of two maternally expressed negative regulators of seed dormancy, DOG1-LIKE 4 (DOGL4) and ALLANTOINASE (ALN) [106,107]. In *A. thaliana*, active DNA demethylation depends on the activity of ROS1, which directly excises methyl-C from DNA [89]. Interestingly, *ros1* mutants are hypersensitive to ABA during early seedling development [108], showing that ROS1-driven DNA demethylation regulates seed dormancy, and the response to ABA by controlling the expression of DOGL4, a negative regulator of seed dormancy and the ABA response. ROS1 appears to antagonize the RdDM pathway, required for the imprinting that underlies the suppression of the paternal allele, by reducing the high level of DNA methylation of the paternal DOGL4 promoter, and consequently the gene repression. While both *dogl4* and *ros1* mutants show a decrease in germination rates and an increase in the response to ABA, overexpression of DOGL4 in *ros1* mutant background reverts the enhanced seed dormancy and ABA hypersensitivity phenotype, which shows that ROS1 regulation of seed dormancy is via DOGL4 [106].

DNA methylation also plays an important role in regulating the preferential maternal *ALN* expression in the seed endosperm. Higher CHH methylation levels in a transposable element present in the 5′ region of *ALN*, targeted by RdDM, correlate with lower expression of the gene. Paternal allele DNA methylation appears to be established during the gametogenesis in sperm cells in this gene [107]. During seed development, cold induces further CHH methylation in this region. This mechanism might allow seeds to store information about cold temperatures and optimize the germination timing. Upon germination, cold-induced DNA methylation is lost, allowing optimal gene expression in the next generation [107]. Further studies are required to understand how the RdDM machinery can discriminate paternal from maternal alleles and maintain the different levels of methylation after fertilization. It is suggested that the RdDM machinery might act preferentially at methylated DNA or that secondary imprinting such as histone modification might direct RdDM to the paternal allele [107].

During germination, although CG and CHG methylations remain stable, which indicates that MET1 and CMT3 are active in maintenance, there is a drop in CHH methylation levels. There is, however, no decrease in siRNAs for these hypomethylated loci nor in RdDM machinery components abundance, which points to demethylation causing the drop [91,95]. DNA demethylases are weakly or not expressed during germination, and demethylases mutants show no significant changes in DNA methylation when compared to wild type [91]. This suggests that passive demethylation may be behind the decrease in DNA methylation levels during the process of germination, although further studies are still needed to support this assertion. More than 12,000 differential methylated regions have been found in analyses upon *A. thaliana* seed germination, with a different degree of methylation in promoters of over 1800 differently expressed genes [95]. This implies that the dynamics of DNA methylation can regulate germination-related gene expression. Nevertheless, further studies on the specific set of genes regulated by changes in DNA methylation levels are required to understand the importance of this epigenetic regulation.

4. Concluding Remarks and Future Perspectives

Although the loss of seed dormancy through AR is a fact observed in a good number of seeds, the molecular mechanisms that govern the AR process are not entirely known. It is widely accepted that oxidation caused by ROS is the main mechanism leading the AR process. Oxidation can be a selective mechanism for degrading specific mRNAs related to proteins that inhibit germination or promote dormancy. The recruitment of mRNAs by RNA binding proteins (RBPs) that protect them during the storage period has been also proposed as a selective mechanism. These results indicate the existence of a specific mechanism that selects long-lived mRNAs essential for the progression of seed germination (Figure 1). It is also proposed that this selective mechanism is susceptible to environmental changes [34]. However, a more detailed molecular scenery is needed to establish the specific players (RBPs) controlling mRNA stability, and to know if other mRNA features are determining its stability upon seed dormancy. Apart from oxidation, other mRNA chemical modifications have been described, such as the methylation of adenosine (m^6A) and cytosine (m^5C) that could modulate transcript stability. In *A. thaliana*, the RNA Adenosine Methylase MTA has been found to be essential for embryo development [80]. New studies approaching this issue are needed to know if this interesting and emerging level to control expression could affect seed dormancy and AR.

DNA methylation is a broadly known epigenetic mechanism controlling gene expression. During *A. thaliana* embryogenesis, there is a global increase in CHH-context methylation, especially at TEs, where both RdDM and CMT2 pathways are involved. Global DNA methylation levels remain stable during seed dormancy and show a substantial loss during germination. This decrease in DNA methylation seems to be mainly due to passive demethylation through subsequent rounds of DNA replication without the maintenance of the methylation pattern. Interestingly, a self-reinforcing loop of repressive epigenetic marks seems to control *DOG1* expression during seed dormancy. The relevance of genomic imprinting, through DNA-methylation, in regulating the maternal expression of *DOG4* and *ALN*, negative regulators of seed dormancy, has been reported in the endosperm of *A. thaliana* seeds. However, the presence of specific DNA methylation marks associated with dormancy or germination transcriptomes remains to be elucidated.

The release of dormancy through AR is considered a major factor controlling soil seed germination in areas where the climate is generally warm and summer is dry, such as occurs in deserts and Mediterranean regions [109,110]. These territories occupy a significant part of the earth: deserts cover over 25% of the land areas, and the five Mediterranean-climate regions more than 2%, where the vascular plant flora comprises approx. 20% of the plant species in the world [111]. From an ecological point of view, the AR requisite avoids seed germinating when a high probability of seedling death exists, due to the elevated temperatures and low rainfalls during the summer [109]. Thus, the AR process can guarantee the species survival and ecosystem biodiversity, highlighting its ecological significance. The molecular mechanisms governing dormancy release through the AR process are not completely known, and the convergent pathways with the stratification process are still undiscovered. Moreover, most of the studies on AR have been carried out in *A. thaliana* ecotypes. This has paved the way to understand this fundamental process of seed physiology, but more knowledge in other species is required. In conclusion, new research about how ROS and nucleic acid modifications interact with other dormancy pathways (i.e., stratification) in *A. thaliana* and the other species would shed light to the uncompleted landscape of dormancy and AR. This will provide new tools for solving relevant agricultural problems such as pre-harvest sprouting.

Author Contributions: R.I.-F. has conceived the main idea, edited and reviewed all manuscript. R.I.-F., K.K.-G., E.C., and N.C.-B. have written the review. K.K.-G. has performed the images. All authors have read and agreed with the final manuscript.

Funding: R. I-F has received funding from "Universidad Politécnica de Madrid" and "Comunidad de Madrid" (Ref.: APOYO-JOVENES-MT0KUE-25-MKITFX).

Acknowledgments: We thank to Pilar Carbonero for the critical reading of the manuscript.

Conflicts of Interest: The authors declare that they have no conflict of interest.

References

1. Vicente-Carbajosa, J.; Carbonero, P. Seed maturation: Developing an intrusive phase to accomplish a quiescent state. *Int. J. Dev. Biol.* **2005**, *49*, 645–651. [CrossRef]

2. Cadman, C.S.C.; Toorop, P.E.; Hilhorst, H.W.M.; Finch-Savage, W.E. Gene expression profiles of Arabidopsis Cvi seeds during dormancy cycling indicate a common underlying dormancy control mechanism. *Plant J.* **2006**, *46*, 805–822. [CrossRef] [PubMed]

3. Holdsworth, M.J.; Bentsink, L.; Soppe, W.J.J. Molecular networks regulating Arabidopsis seed maturation, after-ripening, dormancy and germination. *New Phytol.* **2008**, *179*, 33–54. [CrossRef] [PubMed]

4. Penfield, S. Seed dormancy and germination. *Curr. Biol.* **2017**, *27*, R874–R878. [CrossRef] [PubMed]

5. Gubler, F.; Millar, A.A.; Jacobsen, J.V. Dormancy release, ABA and pre-harvest sprouting. *Curr. Opin. Plant Biol.* **2005**, *8*, 183–187. [CrossRef] [PubMed]

6. Finkelstein, R.; Reeves, W.; Ariizumi, T.; Steber, C. Molecular aspects of seed dormancy. *Annu. Rev. Plant Biol.* **2008**, *59*, 387–415. [CrossRef]

7. Barrero, J.M.; Millar, A.A.; Griffiths, J.; Czechowski, T.; Scheible, W.R.; Udvardi, M.; Reid, J.B.; Ross, J.J.; Jacobsen, J.V.; Gubler, F. Gene expression profiling identifies two regulatory genes controlling dormancy and ABA sensitivity in Arabidopsis seeds. *Plant J.* **2010**, *61*, 611–622. [CrossRef]

8. Abraham, Z.; Iglesias-Fernández, R.; Martínez, M.; Rubio-Somoza, I.; Díaz, I.; Carbonero, P.; Vicente-Carbajosa, J. A developmental switch of gene expression in the barley seed mediated by HvVP1 (Viviparous-1) and HvGAMYB interactions. *Plant Physiol.* **2016**, *170*, 2146–2158. [CrossRef]

9. Shu, K.; Liu, X.; Xie, Q.; He, Z. Two Faces of One Seed: Hormonal regulation of dormancy and germination. *Mol. Plant* **2016**, *9*, 34–45. [CrossRef]

10. Carbonero, P.; Iglesias-Fernández, R.; Vicente-Carbajosa, J. The AFL subfamily of B3 transcription factors: Evolution and function in angiosperm seeds. *J. Exp. Bot.* **2017**, *68*, 871–880. [CrossRef]

11. Nonogaki, H. The long-standing paradox of seed dormancy unfolded? *Trends Plant Sci.* **2019**, *24*, 989–998. [CrossRef] [PubMed]

12. Nakabayashi, K.; Bartsch, M.; Xiang, Y.; Miatton, E.; Pellengahr, S.; Yano, R.; Seo, M.; Soppe, W.J.J. The time required for dormancy release in Arabidopsis is determined by DELAY OF GERMINATION1 protein levels in freshly harvested seeds. *Plant Cell* **2012**, *24*, 2826–2838. [CrossRef] [PubMed]

13. Alonso-Blanco, C.; Bentsink, L.; Hanhart, C.J.; Blankestijn-de Vries, H.; Koornneef, M. Analysis of natural allelic variation at seed dormancy loci of *Arabidopsis thaliana*. *Genetics* **2003**, *164*, 711–729. [CrossRef] [PubMed]

14. Bentsink, L.; Jowett, J.; Hanhart, C.J.; Koornneef, M. Cloning of DOG1, a quantitative trait locus controlling seed dormancy in Arabidopsis. *Proc. Natl. Acad. Sci. USA* **2006**, *103*, 17042–17047. [CrossRef] [PubMed]

15. Née, G.; Xiang, Y.; Soppe, W.J.J. The release of dormancy, a wake-up call for seeds to germinate. *Curr. Opin. Plant Biol.* **2017**, *35*, 8–14. [CrossRef] [PubMed]

16. Nishimura, N.; Tsuchiya, W.; Moresco, J.J.; Hayashi, Y.; Satoh, K.; Kaiwa, N.; Irisa, T.; Kinoshita, T.; Schroeder, J.I.; Yates, J.R.; et al. Control of seed dormancy and germination by DOG1-AHG1 PP2C phosphatase complex via binding to heme. *Nat. Commun.* **2018**, *9*. [CrossRef]

17. Nonogaki, H. A repressor complex silencing ABA signaling in seeds? *J. Exp. Bot.* **2020**. [CrossRef]

18. Carrillo-Barral, N.; del Carmen Rodríguez-Gacio, M.; Matilla, A.J. Delay of Germination-1 (DOG1): A key to understanding seed dormancy. *Plants* **2020**, *9*, 480. [CrossRef]

19. Baskin, J.M.; Baskin, C.C. A classification system for seed dormancy. *Seed Sci. Res.* **2004**, *14*, 1–16. [CrossRef]

20. Iglesias-Fernández, R.; Matilla, A. After-ripening alters the gene expression pattern of oxidases involved in the ethylene and gibberellin pathways during early imbibition of Sisymbrium officinale L. seeds. *J. Exp. Bot.* **2009**, *60*, 1645–1661. [CrossRef]

21. Chahtane, H.; Kim, W.; Lopez-Molina, L. Primary seed dormancy: A temporally multilayered riddle waiting to be unlocked. *J. Exp. Bot.* **2017**, *68*, 857–869. [CrossRef] [PubMed]

22. Buijs, G.; Vogelzang, A.; Nijveen, H.; Bentsink, L. Dormancy cycling: Translation-related transcripts are the main difference between dormant and non-dormant seeds in the field. *Plant J.* **2019**. [CrossRef] [PubMed]

23. Donohue, K.; Dorn, L.; Griffith, C.; Kim, E.S.; Aguilera, A.; Polisetty, C.R.; Schmitt, J. The evolutionary ecology of seed germination of *Arabidopsis thaliana*: Variable natural selection on germination timing. *Evolution* **2005**, *59*, 758–770. [CrossRef] [PubMed]

24. Schmuths, H.; BACHMANN, K.; WEBER, W.E.; HORRES, R.; HOFFMANN, M.H. Effects of preconditioning and temperature during germination of 73 natural accessions of *Arabidopsis thaliana*. *Ann. Bot.* **2006**, *97*, 623–634. [CrossRef]

25. Nakabayashi, K.; Okamoto, M.; Koshiba, T.; Kamiya, Y.; Nambara, E. Genome-wide profiling of stored mRNA in *Arabidopsis thaliana* seed germination: Epigenetic and genetic regulation of transcription in seed. *Plant J.* **2005**, *41*, 697–709. [CrossRef]

26. Carrera, E.; Holman, T.; Medhurst, A.; Dietrich, D.; Footitt, S.; Theodoulou, F.L.; Holdsworth, M.J. Seed after-ripening is a discrete developmental pathway associated with specific gene networks in Arabidopsis. *Plant J.* **2008**, *53*, 214–224. [CrossRef]

27. Carrillo-Barral, N.; Matilla, A.J.; Iglesias-Fernández, R.; del Carmen Rodríguez-Gacio, M. Nitrate-induced early transcriptional changes during imbibition in non-after-ripenedSisymbrium officinaleseeds. *Physiol. Plant.* **2013**, *148*, 560–573. [CrossRef]

28. Carrillo-Barral, N.; Matilla, A.J.; del Carmen Rodríguez-Gacio, M.; Iglesias-Fernández, R. Nitrate affects sensu-stricto germination of after-ripened *Sisymbrium officinale* seeds by modifying expression of *SoNCED5*, *SoCYP707A2* and *SoGA3ox2* genes. *Plant Sci.* **2014**, *217–218*, 99–108. [CrossRef]

29. Dekkers, B.J.W.; Pearce, S.P.; van Bolderen-Veldkamp, R.P.M.; Holdsworth, M.J.; Bentsink, L. Dormant and after-Ripened *Arabidopsis thaliana* seeds are distinguished by early transcriptional differences in the imbibed state. *Front. Plant Sci.* **2016**, *7*. [CrossRef]

30. Bazin, J.; Langlade, N.; Vincourt, P.; Arribat, S.; Balzergue, S.; El-Maarouf-Bouteau, H.; Bailly, C. Targeted mRNA oxidation regulates sunflower seed dormancy alleviation during dry after-ripening. *Plant Cell* **2011**, *23*, 2196–2208. [CrossRef]

31. Bailly, C. The signalling role of ROS in the regulation of seed germination and dormancy. *Biochem. J.* **2019**, *476*, 3019–3032. [CrossRef] [PubMed]

32. Finch-Savage, W.E.; Cadman, C.S.C.; Toorop, P.E.; Lynn, J.R.; Hilhorst, H.W.M. Seed dormancy release in Arabidopsis Cvi by dry after-ripening, low temperature, nitrate and light shows common quantitative patterns of gene expression directed by environmentally specific sensing. *Plant J.* **2007**, *51*, 60–78. [CrossRef] [PubMed]

33. Sano, N.; Permana, H.; Kumada, R.; Shinozaki, Y.; Tanabata, T.; Yamada, T.; Hirasawa, T.; Kanekatsu, M. Proteomic analysis of embryonic proteins synthesized from long-Lived mRNAs during germination of rice seeds. *Plant Cell Physiol.* **2012**, *53*, 687–698. [CrossRef] [PubMed]

34. Galland, M.; Huguet, R.; Arc, E.; Cueff, G.; Job, D.; Rajjou, L. Dynamic proteomics emphasizes the importance of selective mRNA translation and protein turnover during Arabidopsis seed germination. *Mol. Cell. Proteom.* **2014**, *13*, 252–268. [CrossRef]

35. Sano, N.; Rajjou, L.; North, H.M. Lost in translation: Physiological roles of stored mRNAs in seed germination. *Plants* **2020**, *9*, 347. [CrossRef]

36. Xiao, W.; Custard, R.D.; Brown, R.C.; Lemmon, B.E.; Harada, J.J.; Goldberg, R.B.; Fischer, R.L. DNA methylation is critical for Arabidopsis embryogenesis and seed viability. *Plant Cell* **2006**, *18*, 805–814. [CrossRef]

37. Jeevan Kumar, S.P.; Rajendra Prasad, S.; Banerjee, R.; Thammineni, C. Seed birth to death: Dual functions of reactive oxygen species in seed physiology. *Ann. Bot.* **2015**, *116*, 663–668. [CrossRef]

38. Buijs, G.; Kodde, J.; Groot, S.P.C.; Bentsink, L. Seed dormancy release accelerated by elevated partial pressure of oxygen is associated with DOG loci. *J. Exp. Bot.* **2018**, *69*, 3601–3608. [CrossRef]

39. Kurek, K.; Plitta-Michalak, B.; Ratajczak, E. Reactive Oxygen species as potential drivers of the seed aging process. *Plants* **2019**, *8*, 174. [CrossRef]

40. Oracz, K.; Bouteau, H.E.-M.; Farrant, J.M.; Cooper, K.; Belghazi, M.; Job, C.; Job, D.; Corbineau, F.; Bailly, C. ROS production and protein oxidation as a novel mechanism for seed dormancy alleviation. *Plant J.* **2007**, *50*, 452–465. [CrossRef]

41. Müller, K.; Carstens, A.C.; Linkies, A.; Torres, M.A.; Leubner-Metzger, G. The NADPH-oxidase AtrbohB plays a role in Arabidopsis seed after-ripening. *New Phytol.* **2009**, *184*, 885–897. [CrossRef] [PubMed]

42. Ebone, L.A.; Caverzan, A.; Chavarria, G. Physiologic alterations in orthodox seeds due to deterioration processes. *Plant Physiol. Biochem.* **2019**, *145*, 34–42. [CrossRef] [PubMed]

43. De Giorgi, J.; Piskurewicz, U.; Loubery, S.; Utz-Pugin, A.; Bailly, C.; Mène-Saffrané, L.; Lopez-Molina, L. An endosperm-associated cuticle is required for Arabidopsis seed viability, dormancy and early control of germination. *PLOS Genet.* **2015**, *11*, e1005708. [CrossRef] [PubMed]

44. Nakabayashi, K.; Bartsch, M.; Ding, J.; Soppe, W.J.J. Seed dormancy in Arabidopsis requires self-binding ability of DOG1 Protein and the presence of multiple isoforms generated by alternative splicing. *PLOS Genet.* **2015**, *11*, e1005737. [CrossRef]

45. Liu, Y.; Ye, N.; Liu, R.; Chen, M.; Zhang, J. H_2O_2 mediates the regulation of ABA catabolism and GA biosynthesis in Arabidopsis seed dormancy and germination. *J. Exp. Bot.* **2010**, *61*, 2979–2990. [CrossRef]

46. Barba-Espín, G.; Díaz-Vivancos, P.; Job, D.; Belghazi, M.; Job, C.; Hernández, J.A. Understanding the role of H_2O_2 during pea seed germination: A combined proteomic and hormone profiling approach. *Plant. Cell Environ.* **2011**, *34*, 1907–1919. [CrossRef]

47. Yu, Y.; Zhen, S.; Wang, S.; Wang, Y.; Cao, H.; Zhang, Y.; Li, J.; Yan, Y. Comparative transcriptome analysis of wheat embryo and endosperm responses to ABA and H_2O_2 stresses during seed germination. *BMC Genom.* **2016**, *17*. [CrossRef]

48. Ishibashi, Y.; Aoki, N.; Kasa, S.; Sakamoto, M.; Kai, K.; Tomokiyo, R.; Watabe, G.; Yuasa, T.; Iwaya-Inoue, M. The interrelationship between abscisic acid and reactive oxygen species plays a key role in barley seed dormancy and germination. *Front. Plant Sci.* **2017**, *8*. [CrossRef]

49. Leymarie, J.; Vitkauskaité, G.; Hoang, H.H.; Gendreau, E.; Chazoule, V.; Meimoun, P.; Corbineau, F.; El-Maarouf-Bouteau, H.; Bailly, C. Role of reactive oxygen species in the regulation of Arabidopsis seed dormancy. *Plant Cell Physiol.* **2012**, *53*, 96–106. [CrossRef]

50. Morscher, F.; Kranner, I.; Arc, E.; Bailly, C.; Roach, T. Glutathione redox state, tocochromanols, fatty acids, antioxidant enzymes and protein carbonylation in sunflower seed embryos associated with after-ripening and ageing. *Ann. Bot.* **2015**, *116*, 669–678. [CrossRef]

51. Vigliocco, A.; Del Bel, Z.; Pérez-Chaca, M.V.; Molina, A.; Zirulnik, F.; Andrade, A.M.; Alemano, S. Spatiotemporal variations in salicylic acid and hydrogen peroxide in sunflower seeds during transition from dormancy to germination. *Physiol. Plant.* **2019**. [CrossRef] [PubMed]

52. Müller, K.; Linkies, A.; Vreeburg, R.A.M.; Fry, S.C.; Krieger-Liszkay, A.; Leubner-Metzger, G. In vivo cell wall loosening by hydroxyl radicals during cress seed germination and elongation growth. *Plant Physiol.* **2009**, *150*, 1855–1865. [CrossRef] [PubMed]

53. Oracz, K.; Voegele, A.; Tarkowská, D.; Jacquemoud, D.; Turečková, V.; Urbanová, T.; Strnad, M.; Sliwinska, E.; Leubner-Metzger, G. Myrigalone A inhibits *Lepidium sativum* seed germination by interference with gibberellin metabolism and apoplastic superoxide production required for embryo extension growth and endosperm rupture. *Plant Cell Physiol.* **2012**, *53*, 81–95. [CrossRef] [PubMed]

54. Waszczak, C.; Carmody, M.; Kangasjärvi, J. Reactive Oxygen Species in plant signaling. *Annu. Rev. Plant Biol.* **2018**, *69*, 209–236. [CrossRef]

55. Oracz, K.; El-Maarouf-Bouteau, H.; Kranner, I.; Bogatek, R.; Corbineau, F.; Bailly, C. The mechanisms involved in seed dormancy alleviation by hydrogen cyanide unravel the role of eactive Oxygen Species as key factors of cellular signaling during germination. *Plant Physiol.* **2009**, *150*, 494–505. [CrossRef]

56. Bahin, E.; Bailly, C.; Sotta, B.; Kranner, I.; Corbineau, F.; Leymarie, J. Crosstalk between reactive oxygen species and hormonal signalling pathways regulates grain dormancy in barley. *Plant Cell Environ.* **2011**, *34*, 980–993. [CrossRef]

57. Kaur, K.; Zhawar, V.K. Antioxidant potential of fresh and after-ripened dry embryos of two wheat cultivars contrasting in drought tolerance. *Indian J. Plant Physiol.* **2016**, *21*, 350–354. [CrossRef]

58. Farooq, M.A.; Niazi, A.K.; Akhtar, J.; Saifullah Farooq, M.; Souri, Z.; Karimi, N.; Rengel, Z. Acquiring control: The evolution of ROS-Induced oxidative stress and redox signaling pathways in plant stress responses. *Plant Physiol. Biochem.* **2019**, *141*, 353–369. [CrossRef]

59. Serrato, A.J.; Crespo, J.L.; Florencio, F.J.; Cejudo, F.J. Characterization of two thioredoxins h with predominant localization in the nucleus of aleurone and scutellum cells of germinating wheat seeds. *Plant Mol. Biol.* **2001**, *46*, 361–371. [CrossRef]

60. Ortiz-Espín, A.; Iglesias-Fernández, R.; Calderón, A.; Carbonero, P.; Sevilla, F.; Jiménez, A. Mitochondrial *AtTrxo1* is transcriptionally regulated by AtbZIP9 and AtAZF2 and affects seed germination under saline conditions. *J. Exp. Bot.* **2017**, *68*, 1025–1038. [CrossRef]

61. Xu, F.; Tang, J.; Gao, S.; Cheng, X.; Du, L.; Chu, C. Control of rice pre-harvest sprouting by glutaredoxin-mediated abscisic acid signaling. *Plant J.* **2019**, *100*, 1036–1051. [CrossRef] [PubMed]

62. Chmielowska-Bąk, J.; Arasimowicz-Jelonek, M.; Deckert, J. In search of the mRNA modification landscape in plants. *BMC Plant Biol.* **2019**, *19*. [CrossRef] [PubMed]

63. Simms, C.L.; Hudson, B.H.; Mosior, J.W.; Rangwala, A.S.; Zaher, H.S. An active role for the ribosome in determining the fate of oxidized mRNA. *Cell Rep.* **2014**, *9*, 1256–1264. [CrossRef] [PubMed]

64. Taddei, F.; Hayakawa, H.; Bouton, M.-F.; Cirinesi, A.-M.; Matic, I.; Sekiguchi, M.; Radman, M. Counteraction by MutT protein of transcriptional errors caused by oxidative damage. *Science* **1997**, *278*, 128–130. [CrossRef] [PubMed]

65. Gao, F.; Rampitsch, C.; Chitnis, V.R.; Humphreys, G.D.; Jordan, M.C.; Ayele, B.T. Integrated analysis of seed proteome and mRNA oxidation reveals distinct post-transcriptional features regulating dormancy in wheat (*Triticum aestivum* L.). *Plant Biotechnol. J.* **2013**, *11*, 921–932. [CrossRef]

66. Gao, F.; Jordan, M.C.; Ayele, B.T. Microarray dataset of after-ripening induced mRNA oxidation in wheat seeds. *Data Br.* **2018**, *21*, 852–855. [CrossRef]

67. Nelson, S.K.; Ariizumi, T.; Steber, C.M. Biology in the dry seed: Transcriptome changes associated with dry seed dormancy and dormancy loss in the arabidopsis GA-insensitive sleepy1-2 mutant. *Front. Plant Sci.* **2017**, *8*. [CrossRef]

68. Cléry, A.; Blatter, M.; Allain, F.H.-T. RNA recognition motifs: Boring? Not quite. *Curr. Opin. Struct. Biol.* **2008**, *18*, 290–298. [CrossRef]

69. Chai, Q.; Singh, B.; Peisker, K.; Metzendorf, N.; Ge, X.; Dasgupta, S.; Sanyal, S. Organization of ribosomes and nucleoids in *Escherichia coli* cells during growth and in quiescence. *J. Biol. Chem.* **2014**, *289*, 11342–11352. [CrossRef]

70. Bai, B.; van der Horst, S.; Cordewener, J.H.G.; America, T.A.H.P.; Hanson, J.; Bentsink, L. Seed-stored mRNAs that are specifically associated to monosomes are translationally regulated during germination. *Plant Physiol.* **2020**, *182*, 378–392. [CrossRef]

71. El-Maarouf-Bouteau, H.; Meimoun, P.; Job, C.; Job, D.; Bailly, C. Role of protein and mRNA oxidation in seed dormancy and germination. *Front. Plant Sci.* **2013**, *4*. [CrossRef] [PubMed]

72. Chen, H.; Chu, P.; Zhou, Y.; Li, Y.; Liu, J.; Ding, Y.; Tsang, E.W.T.; Jiang, L.; Wu, K.; Huang, S. Overexpression of AtOGG1, a DNA glycosylase/AP lyase, enhances seed longevity and abiotic stress tolerance in Arabidopsis. *J. Exp. Bot.* **2012**, *63*, 4107–4121. [CrossRef] [PubMed]

73. Madugundu, G.S.; Cadet, J.; Wagner, J.R. Hydroxyl-radical-induced oxidation of 5-methylcytosine in isolated and cellular DNA. *Nucleic Acids Res.* **2014**, *42*, 7450–7460. [CrossRef] [PubMed]

74. Michalak, M.; Barciszewska, M.Z.; Barciszewski, J.; Plitta, B.P.; Chmielarz, P. Global Changes in DNA methylation in seeds and seedlings of *Pyrus communis* after Seed desiccation and storage. *PLoS ONE* **2013**, *8*, e70693. [CrossRef]

75. Michalak, M.; Plitta-Michalak, B.P.; Naskręt-Barciszewska, M.; Barciszewski, J.; Bujarska-Borkowska, B.; Chmielarz, P. Global 5-methylcytosine alterations in DNA during ageing of *Quercus robur* seeds. *Ann. Bot.* **2015**, *116*, 369–376. [CrossRef]

76. Shen, L.; Liang, Z.; Wong, C.E.; Yu, H. Messenger RNA modifications in plants. *Trends Plant Sci.* **2019**, *24*, 328–341. [CrossRef]

77. Cui, X.; Liang, Z.; Shen, L.; Zhang, Q.; Bao, S.; Geng, Y.; Zhang, B.; Leo, V.; Vardy, L.A.; Lu, T.; et al. 5-Methylcytosine RNA methylation in *Arabidopsis thaliana*. *Mol. Plant* **2017**, *10*, 1387–1399. [CrossRef]

78. Vandivier, L.E.; Campos, R.; Kuksa, P.P.; Silverman, I.M.; Wang, L.-S.; Gregory, B.D. Chemical modifications mark alternatively spliced and uncapped messenger RNAs in Arabidopsis. *Plant Cell* **2015**, *27*, 3024–3037. [CrossRef]

79. Fray, R.G.; Simpson, G.G. The Arabidopsis epitranscriptome. *Curr. Opin. Plant Biol.* **2015**, *27*, 17–21. [CrossRef]

80. Zhong, S.; Li, H.; Bodi, Z.; Button, J.; Vespa, L.; Herzog, M.; Fray, R.G. MTA Iis an Arabidopsis Messenger RNA adenosine methylase and interacts with a homolog of a sex-specific splicing factor. *Plant Cell* **2008**, *20*, 1278–1288. [CrossRef]

81. Bodi, Z.; Zhong, S.; Mehra, S.; Song, J.; Graham, N.; Li, H.; May, S.; Fray, R.G. Adenosine methylation in Arabidopsis mRNA is associated with the 3′ end and reduced levels cause developmental defects. *Front. Plant Sci.* **2012**, *3*. [CrossRef]

82. Diez, C.M.; Roessler, K.; Gaut, B.S. Epigenetics and plant genome evolution. *Curr. Opin. Plant Biol.* **2014**, *18*, 1–8. [CrossRef]

83. Zhang, X.; Yazaki, J.; Sundaresan, A.; Cokus, S.; Chan, S.W.L.; Chen, H.; Henderson, I.R.; Shinn, P.; Pellegrini, M.; Jacobsen, S.E.; et al. Genome-wide high-resolution mapping and functional analysis of DNA methylation in Arabidopsis. *Cell* **2006**, *126*, 1189–1201. [CrossRef]

84. Bewick, A.J.; Schmitz, R.J. Gene body DNA methylation in plants. *Curr. Opin. Plant Biol.* **2017**, *36*, 103–110. [CrossRef]

85. Zilberman, D.; Gehring, M.; Tran, R.K.; Ballinger, T.; Henikoff, S. Genome-wide analysis of *Arabidopsis thaliana* DNA methylation uncovers an interdependence between methylation and transcription. *Nat. Genet.* **2007**, *39*, 61–69. [CrossRef]

86. Law, J.A.; Jacobsen, S.E. Establishing, maintaining and modifying DNA methylation patterns in plants and animals. *Nat. Rev. Genet.* **2010**, *11*, 204–220. [CrossRef] [PubMed]

87. Matzke, M.A.; Kanno, T.; Matzke, A.J.M. RNA-Directed DNA methylation: The evolution of a complex epigenetic pathway in flowering plants. *Annu. Rev. Plant Biol.* **2015**, *66*, 243–267. [CrossRef] [PubMed]

88. Zhang, H.; Lang, Z.; Zhu, J.-K. Dynamics and function of DNA methylation in plants. *Nat. Rev. Mol. Cell Biol.* **2018**, *19*, 489–506. [CrossRef] [PubMed]

89. Parrilla-Doblas, J.T.; Roldán-Arjona, T.; Ariza, R.R.; Córdoba-Cañero, D. Active DNA Demethylation in plants. *Int. J. Mol. Sci.* **2019**, *20*, 4683. [CrossRef] [PubMed]

90. van Zanten, M.; Koini, M.A.; Geyer, R.; Liu, Y.; Brambilla, V.; Bartels, D.; Koornneef, M.; Fransz, P.; Soppe, W.J.J. Seed maturation in *Arabidopsis thaliana* is characterized by nuclear size reduction and increased chromatin condensation. *Proc. Natl. Acad. Sci. USA* **2011**, *108*, 20219–20224. [CrossRef]

91. Kawakatsu, T.; Nery, J.R.; Castanon, R.; Ecker, J.R. Dynamic DNA methylation reconfiguration during seed development and germination. *Genome Biol.* **2017**, *18*, 171. [CrossRef] [PubMed]

92. An, Y.C.; Goettel, W.; Han, Q.; Bartels, A.; Liu, Z.; Xiao, W. Dynamic changes of genome-wide DNA methylation during soybean seed development. *Sci. Rep.* **2017**, *7*. [CrossRef] [PubMed]

93. Bouyer, D.; Kramdi, A.; Kassam, M.; Heese, M.; Schnittger, A.; Roudier, F.; Colot, V. DNA methylation dynamics during early plant life. *Genome Biol.* **2017**, *18*, 179. [CrossRef] [PubMed]

94. Lin, J.-Y.; Le, B.H.; Chen, M.; Henry, K.F.; Hur, J.; Hsieh, T.-F.; Chen, P.-Y.; Pelletier, J.M.; Pellegrini, M.; Fischer, R.L.; et al. Similarity between soybean and Arabidopsis seed methylomes and loss of non-CG methylation does not affect seed development. *Proc. Natl. Acad. Sci. USA* **2017**, *114*, E9730–E9739. [CrossRef]

95. Narsai, R.; Gouil, Q.; Secco, D.; Srivastava, A.; Karpievitch, Y.V.; Liew, L.C.; Lister, R.; Lewsey, M.G.; Whelan, J. Extensive transcriptomic and epigenomic remodelling occurs during *Arabidopsis thaliana* germination. *Genome Biol.* **2017**, *18*. [CrossRef]

96. Choi, Y.; Gehring, M.; Johnson, L.; Hannon, M.; Harada, J.J.; Goldberg, R.B.; Jacobsen, S.E.; Fischer, R.L. DEMETER, a DNA glycosylase domain protein, is required for endosperm gene imprinting and seed viability in Arabidopsis. *Cell* **2002**, *110*, 33–42. [CrossRef]

97. Xiao, W.; Brown, R.C.; Lemmon, B.E.; Harada, J.J.; Goldberg, R.B.; Fischer, R.L. Regulation of seed size by hypomethylation of maternal and paternal genomes. *Plant Physiol.* **2006**, *142*, 1160–1168. [CrossRef]

98. Pettersen, E.F.; Goddard, T.D.; Huang, C.C.; Couch, G.S.; Greenblatt, D.M.; Meng, E.C.; Ferrin, T.E. UCSF Chimera—A visualization system for exploratory research and analysis. *J. Comput. Chem.* **2004**, *25*, 1605–1612. [CrossRef]

99. Li, P.; Ni, H.; Ying, S.; Wei, J.; Hu, X. Teaching an old dog a new trick: Multifaceted strategies to control primary seed germination by DELAY OF GERMINATION 1 (DOG1). *Phyton (B. Aires)* **2020**, *89*, 1–12. [CrossRef]

100. Zheng, J.; Chen, F.; Wang, Z.; Cao, H.; Li, X.; Deng, X.; Soppe, W.J.J.; Li, Y.; Liu, Y. A novel role for histone methyltransferase KYP/SUVH4 in the control of Arabidopsis primary seed dormancy. *New Phytol.* **2012**, *193*, 605–616. [CrossRef]

101. Singh, M.; Singh, J. Seed development-related expression of ARGONAUTE4_9 class of genes in barley: Possible role in seed dormancy. *Euphytica* **2012**, *188*, 123–129. [CrossRef]

102. Singh, M.; Singh, S.; Randhawa, H.; Singh, J. Polymorphic homoeolog of key gene of RdDM pathway, ARGONAUTE4_9 class is associated with pre-harvest sprouting in Wheat (*Triticum aestivum* L.). *PLoS ONE* **2013**, *8*. [CrossRef] [PubMed]

103. Zha, P.; Liu, S.; Li, Y.; Ma, T.; Yang, L.; Jing, Y.; Lin, R. The evening complex and the chromatin-remodeling factor PICKLE coordinately control seed dormancy by directly repressing *DOG1* in Arabidopsis. *Plant Commun.* **2020**, *1*, 100011. [CrossRef]

104. Köhler, C.; Wolff, P.; Spillane, C. Epigenetic mechanisms underlying genomic imprinting in plants. *Annu. Rev. Plant Biol.* **2012**, *63*, 331–352. [CrossRef] [PubMed]

105. Piskurewicz, U.; Iwasaki, M.; Susaki, D.; Megies, C.; Kinoshita, T.; Lopez-Molina, L. Dormancy-specific imprinting underlies maternal inheritance of seed dormancy in *Arabidopsis thaliana*. *eLife* **2016**. [CrossRef] [PubMed]

106. Zhu, H.; Xie, W.; Xu, D.; Miki, D.; Tang, K.; Huang, C.-F.; Zhu, J.-K. DNA demethylase ROS1 negatively regulates the imprinting of *DOGL4* and seed dormancy in *Arabidopsis thaliana*. *Proc. Natl. Acad. Sci. USA* **2018**, *115*, E9962–E9970. [CrossRef]

107. Iwasaki, M.; Hyvärinen, L.; Piskurewicz, U.; Lopez-Molina, L. Non-canonical RNA-directed DNA methylation participates in maternal and environmental control of seed dormancy. *eLife* **2019**, *8*. [CrossRef]

108. Kim, J.-S.; Lim, J.Y.; Shin, H.; Kim, B.-G.; Yoo, S.-D.; Kim, W.T.; Huh, J.H. ROS1-dependent DNA demethylation is required for ABA-inducible NIC3 expression. *Plant Physiol.* **2019**, *179*, 1810–1821. [CrossRef]

109. Iglesias-Fernández, R.; Del Carmen Rodrguez-Gacio, M.; Matilla, A.J. Progress in research on dry afterripening. *Seed Sci. Res.* **2011**, *21*, 69–80. [CrossRef]

110. Baskin, C.C.; Baskin, J.M. *Seeds: Ecology, Biogeography, and, Evolution of Dormancy and Germination*, 2nd ed.; Academic Press: Cambridge, MA, USA, 2014; pp. 79–117.

111. Klausmeyer, K.R.; Shaw, M.R. Climate change, habitat loss, protected areas and the climate adaptation potential of species in mediterranean ecosystems worldwide. *PLoS ONE* **2009**, *4*. [CrossRef]

Review

Delay of Germination-1 (DOG1): A Key to Understanding Seed Dormancy

Néstor Carrillo-Barral [1], María del Carmen Rodríguez-Gacio [2] and Angel Jesús Matilla [2,*]

[1] Departamento de Biología, Facultad de Ciencias, Universidad de A Coruña, Campus Zapateira, 15071-A Coruña, Spain; n.carrillo@udc.es
[2] Departamento de Biología Funcional (Área Fisiología Vegetal), Facultad de Farmacia, Universidad de Santiago de Compostela, 15782 Santiago de Compostela, Spain; mdelcarmen.rodriguez.gacio@usc.es
* Correspondence: angeljesus.matilla@usc.es; Tel.: +34-981-563-100

Received: 21 February 2020; Accepted: 3 April 2020; Published: 9 April 2020

Abstract: DELAY OF GERMINATION-1 (DOG1), is a master regulator of primary dormancy (PD) that acts in concert with ABA to delay germination. The ABA and DOG1 signaling pathways converge since DOG1 requires protein phosphatase 2C (PP2C) to control PD. DOG1 enhances ABA signaling through its binding to PP2C ABA HYPERSENSITIVE GERMINATION (AHG1/AHG3). DOG1 suppresses the AHG1 action to enhance ABA sensitivity and impose PD. To carry out this suppression, the formation of DOG1-heme complex is essential. The binding of DOG1-AHG1 to DOG1-Heme is an independent processes but essential for DOG1 function. The quantity of active DOG1 in mature and viable seeds is correlated with the extent of PD. Thus, *dog1* mutant seeds, which have scarce endogenous ABA and high gibberellin (GAs) content, exhibit a non-dormancy phenotype. Despite being studied extensively in recent years, little is known about the molecular mechanism underlying the transcriptional regulation of *DOG1*. However, it is well-known that the physiological function of DOG1 is tightly regulated by a complex array of transformations that include alternative splicing, alternative polyadenylation, histone modifications, and a *cis*-acting antisense non-coding transcript (*asDOG1*). The DOG1 becomes modified (i.e., inactivated) during seed after-ripening (AR), and its levels in viable seeds do not correlate with germination potential. Interestingly, it was recently found that the transcription factor (TF) bZIP67 binds to the *DOG1* promoter. This is required to activate *DOG1* expression leading to enhanced seed dormancy. On the other hand, seed development under low-temperature conditions triggers *DOG1* expression by increasing the expression and abundance of bZIP67. Together, current data indicate that DOG1 function is not strictly limited to PD process, but that it is also required for other facets of seed maturation, in part by also interfering with the ethylene signaling components. Otherwise, since DOG1 also affects other processes such us flowering and drought tolerance, the approaches to understanding its mechanism of action and control are, at this time, still inconclusive.

Keywords: DOG1; seed dormancy; ABA; ethylene; clade-A PP2C phosphatase (AHG1; AHG3); after-ripening; asDOG1; heme-group

---

1. Introduction

The seed is a key entity in the life cycle of higher plants. To this end, it enables and ensures its survival by acquiring the primary dormancy (PD) during the maturation phase. PD is defined as the incapacity of a viable seed to complete germination despite the conditions are favorable [1]. Dormancy is hormonally induced, maintained, and strictly regulated by the modulation of suitable hormonal signaling networks [1]. Seeds can detect spatial and temporal environmental conditions (e.g., temperature, O_2 and light) [1–4]. PD dormancy is a notable agronomic feature. Thus, low levels may produce pre-harvest sprouting, while high levels inhibit the rapid and uniform germination rate. In both cases, there is a reduction of crop production [1,4,5]. PD is established and maintained in the viable dry seed, and throughout several molecular paths (e.g., after-ripening (AR) and exposure to gentle cooling) and can be broken gradually [6,7]. On the other hand, non-dormant seeds of various species have been reported to achieve secondary dormancy (SD) upon exposure to unfavorable conditions for germination [1]. SD occurs essentially after seed dispersal and may be induced by environmental interactions or other special conditions.

DELAY OF GERMINATION1 (*DOG1*), a heme-binding protein, was identified in seeds of *A. thaliana* by using a biparental recombinant inbred population derived from a cross between the deeply dormant ecotype Cape Verde Islands (Cvi-0; high *DOG1* expression) and a weakly dormant ecotype Landsberg *erecta* (Ler-0; low *DOG1* expression). By analyzing this recombinant inbred line (RIL) population, seven dormancy related quantitative trait loci (QTLs) were identified [8,9]. The first of these QTLs cloned was *DOG1* which turned out to have a great impact on PD level (i.e., became the most important regulator of PD in Arabidopsis) [2,9]. *DOG1* is mainly expressed in seeds [10]. Despite being studied extensively over the last fifteen years [10], the precise molecular mechanism underlying the regulation of DOG1 still has gaps [10–12]. Interestingly, Li et al. (2019) provides mechanistic insights into how ethylene (ET) signaling controls PD via DOG1 regulation. Detailed comments on this striking work are included below.

DOG1 is a member of a small gene family. There are five *DOG1-LIKE* (*DOGL*) genes—*DOGL1*, *DOGL2*, *DOGL3*, *DOGL4*, and *DOGL5*—in the Arabidopsis genome [1]. The first four *DOGL* genes are located on chromosome 4 next to each other, while *DOGL5* was found on chromosome 3. The DOG1 family in Arabidopsis has several conserved domains whose functions are still under study [1,5]. DOGL1, DOGL2, and DOGL3 are relatively like DOG1, whereas DOGL4 and DOGL5 show only 28 and 30% similarity with DOG1 in amino acid sequence, respectively [1]. In parallel to Arabidopsis, *DOG1* genes have been found in other species of Brassicaceae [7,11,12] and some of *Lactuca* [13]. The amino acid sequences similarity among the studied dicot DOG1 is high [7,14]. Other authors have described *DOGL* genes in many monocot species, such as *Hordeum vulgare*, *Triticum aestivum*, *Oryza sativum*, *Zea mays*, *Sorghum bicolor*, and *Brachypodium distachyon* [5,14,15] (Figure 1, Figure S1, Table S1). A high number of *AtDOG1* homologues have been described in cereals. DOG1L1-4 in cereals show some functional conservation since they induce dormancy when ectopically expressed in Arabidopsis. However, the amino acid sequences of cereal DOG1L do not have much similarity with those of dicot plants. Although genes with some degree of similarity in amino acid sequence to *AtDOG1* were known to be present in species other than Arabidopsis, because of their low similarity to *DOG1* it remained to be answered whether these genes actually functioned in the regulation of PD in a broad range of species, or only in Arabidopsis and its close relatives [11]. *AtDOG1* and its homologous genes would be good candidates to try to manipulate PD in the near future. It has been found that *DOG1* expression and PD are controlled by a *cis*-acting antisense transcript (*asDOG1*) [16]. As will be described later, *asDOG1* expression is strongly suppressed by both ABA and drought, resulting in the release of antisense-dependent silencing of *DOG1* [16]. Moreover, it was shown that DOG1 is also involved in the seed maturation programme, seed longevity, PD release, and germination timing, this last property being an important adaptive trait controlled by seed dormancy. At this time, the detailed involvement of DOG1 in all these cited programmes is not yet well understood [9–11,13]. However, a notable function of DOG1 may be the induction of PD including the temperature-dependent PD [10,17–22]. Thus, it is known that the lower the seed maturation temperature, the higher the degree of dormancy. Concomitantly, DOG1-mRNA and protein levels also increase [17]. That is, *A. thaliana* accessions located in colder climates tend to initiate DOG1 expression earlier during seed maturation. *DOG1* is involved in the induction of PD in response to cool seed-maturation temperature experienced by the mother plant. Therefore, DOG1 is likely to exhibit environmental sensitivity [19]. Since DOG1 also affects flowering and drought tolerance, DOG1's study becomes much more convoluted [13,22–24].

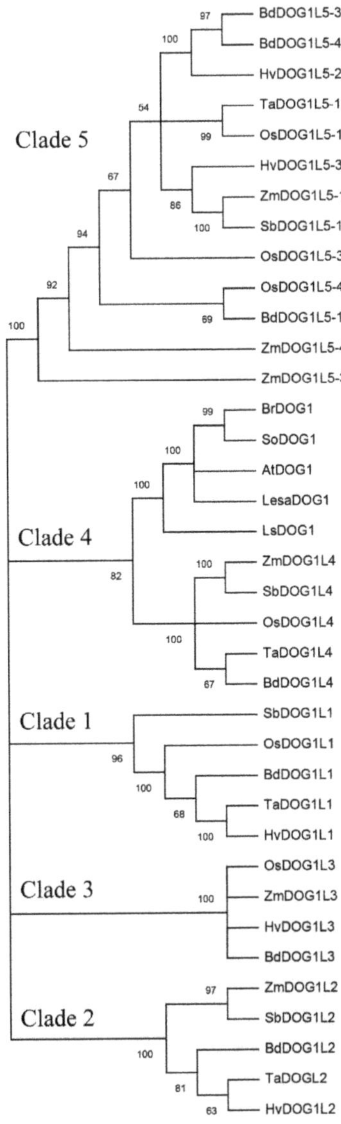

Figure 1. Phylogenetic tree of DOG1. Deduced DOG1 and DOG1-like amino acid sequences of *Arabidopsis thaliana* (AtDOG1, AtDOGL1-5); deduced DOG1 amino acid sequences of *Brassica rapa* (BrDOG1), *Lactuca sativa* (LsDOG1), *Lepidum sativum* (LesaDOG1), *Sisymbrium officinale* (SoDOG1); and the homologs in cereals *Brachypodium distachyon* (BdDOG1L1 to L5), *Hordeum vulgare* (HvDOG1L1, L2, L3, and L5), *Oryza sativa* (OsDOG1L1, L3, L4, and L5), *Sorghum bicolor* (SbDOG1L1 to L5), *Triticum aestivum* (TaDOG1L1, L2, L4, and L5), and *Zea mays* (ZmDOG1L2 to L5) were included. Tree was constructed with the neighbor joining method (1000 bootstrap repetitions) using MEGA X software. Only branches with a value above 50% are shown. An additional table indicates the accession numbers of the sequences included in the tree (Figure S1 and Table S1).

This review focuses on the reasons why DOG1 is being considered as a key signaling molecule to coordinate the seed life and very specifically the acquisition and loss of its PD.

2. DOG1 and Its Vital Functions during the Seed Life Cycle

In the last years, numerous studies have proved that the functions of DOG1 in plants are well conserved. Despite its well-documented implication in PD, DOG1 molecular task still needs much more attention [5,14,25,26].

2.1. DOG1 in Seed Maturation and Dormancy

Seed maturation is a crucial phase of seed development that comprises several developmental processes such as reserve accumulation (e.g., seed storage proteins, SSPs), cessation of embryo growth, acquisition of desiccation tolerance, and PD [27,28]. Although there are many transcription factors (TFs) specifically associated with seed maturation, only ABI3, FUSCA3 (FUS3), LEAFY COTYLEDON 1 (LEC1), and LEC2 have been described as key regulators in Arabidopsis [1,4]. Mutations in these TFs produce alterations in seed maturation, affect accumulation of SSPs, and alter ABA sensibility and PD (e.g., *abi3-lec1* and *abi3-fus3* show an important reduction in SSPs content and present severe viviparism) [1,4,11]. The induction and release of PD depend on environmental conditions (e.g., temperature, light, and cold) and endogenous regulators (e.g., hormones, regulatory proteins, and chromatin status) [20,25,28]. It is noteworthy that the mechanism of endogenous plant hormonal regulation is suggested to be highly conserved in PD and germination processes. In reference to DOG1, its dimerization (Figure 2) is essential for its capability to impose PD [26]. However, it remains unclear how self-dimerization is involved in DOG1 function. Specifically, the expression of *DOG1* is absolutely required for the induction of PD. Conversely, DOG1 is reduced in fully mature dry seed [10,16,23] and *DOG1*-mRNA is nearly undetectable in seedlings [1,16]. It is well-known that during late seed maturation an accumulation of the raffinose family oligosaccharides (RFO), an increase of SSPs, heat-shock proteins (HSPs), and late embryogenesis abundant (LEA) proteins exist [28]. Transcriptomic and metabolomic analyses of *dog1-1* mutants revealed a decrease in HSP, LEA, and RFO. The expression of ABI5 is involved in this decrease [23,29,30]. It is striking that *dog1-1* mutant has reduced PD and reduced longevity. This fact suggests a positive correlation between both processes [9,31]. Several QTLs related to seed longevity have been identified [31]. The QTLs identified for longevity and PD do not necessarily co-locate, proposing that the natural variations of these two characteristics are regulated by different genetic mechanisms. The observation of a negative correlation between PD and longevity strongly suggests that seeds are capable of extending their life span either by PD or by an active longevity mechanism [26]. Furthermore, the analysis of *dog1-1 abi3-1* suggests that DOG1 may control the chlorophyll degradation and SSPs accumulation through an interaction with ABI3. These data also suggest that DOG1 and ABI3 work in parallel to stimulate maturation [23]. Moreover, individual analysis of *dog1* and *dog1 abi3* mutants suggests that DOG1 function is not limited to PD but that it is required for several aspects of seed maturation [23]. Likewise, variations in the *DOG1* expression during the PD seem to be partially due to an epigenetic regulation (i.e., histone methylation) (Figure 2), as will be discussed below.

ABA induces and maintains PD, while GAs releases PD and induces germination. Accordingly, the dormancy status is regulated by the balance ABA: GAs, which will be used for the seed to interpret the environmental conditions and therefore stay dormant or germinate [1,32–35]. Cold temperatures exposure of maternal plants during seed maturation induces an increase in both DOG1 and PD status [20]. Thus, accessions of Arabidopsis from the north of Europe are less dormant and exhibit lower levels of *DOG1* transcripts than the southern ones [17]. On the other hand, the level of dormancy cycling in the field is not quantitatively related to ABA. Accordingly, the analysis of a set of accessions of four distinct geographical areas revealed that *DOG1* contributes to local adaptation [36–38]. DOG1 is important to determine the deep dormancy phase and it could act as part of a thermal-sensing system to affect PD level by altering the ABA sensitivity [34]. A study with buried Arabidopsis seeds showed dynamic changes in SD intensity, which had a strong correlation with *DOG1* expression level [33,36]. Murphey et al. [19] proposed that *DOG1* is a relevant gene in the induction of SD in response to temperatures. In some seeds, DOG1 induces SD because of the warm and prolonged cold stratification in the course of seed imbibition [20,39]. Finch-Savage and Footitt indicated in their recent update on PD [34] that the thermo-inhibition of germination was DOG1 dependent and not reliant on an increased amount of ABA in lab conditions; while DOG1 is of decisive importance to dormancy cycling in field conditions in addition to its importance in determining the extent of PD. Besides, germination only occurs in the light if DOG1 expression is low as a result of chromatin remodeling and the level of active DOG1 protein is reduced [37].

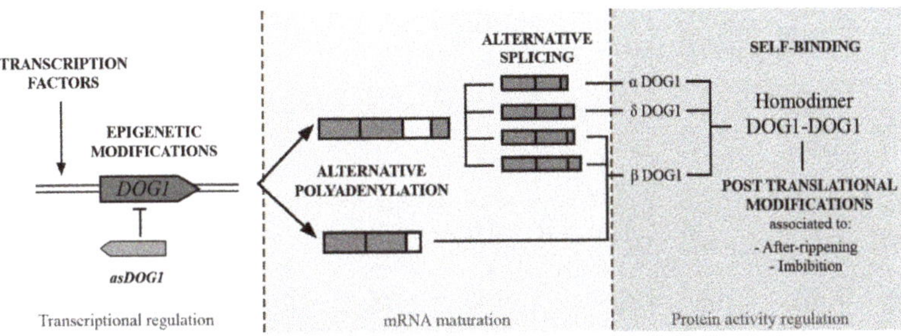

Figure 2. Different molecular mechanisms regulating the gene expression and protein activity of DOG1. The transcription of *DOG1* is regulated by epigenetic modifications and probably by TFs. Additionally, the transcription of a noncoding antisense sequence (asDOG1) acts as a negative regulator of *DOG1* expression. Two different precursor mRNAs are formed due to the existence of two polyadenylation sites in Arabidopsis. The precursor mRNAs are processed to five different mature mRNA by alternative splicing and lately translated to three different protein isoforms (three of the five mRNA encode the same protein isoform). DOG1 binds itself to form homodimers and can suffer post-translational modifications associated to AR and germination processes. However, the specific nature of these modifications is still unknown.

2.2. DOG1 Takes Part in Key Molecular Mechanisms Regulating Seed Dormancy

Recent studies in *Lactuca sativa* and *A. thaliana* reported that DOG1 stimulates the temperature-dependent PD by affecting the levels of determined microRNAs [13]. Thus, DOG1 can regulate PD by means of an influence on generation and/or action and processing of miR156 and miR172 microRNAs [1,13]. miRNAs are endogenous small non-coding RNAs that act as post-transcriptional regulators of gene expression in animals and plants by targeting mRNAs for cleavage or translational repression. miR156 accumulates in high levels during seed development and it is progressively lost during seed storage as a result of RNA oxidation [40,41]. RNA oxidation can be carried out by reactive oxygen species (ROS) generated in dry seeds, particularly by hydroxyl radicals [41]. Seed alleviation of PD during AR is associated with mRNA oxidation which is prevented when seeds are maintained dormant [40]. Therefore, DOG1 may play a role in repositioning the miR156 levels during seed development [22], which would provide a timing mechanism for seed AR via loss of the inhibitory effect of stored miR156 on germination. In short, DOG1 functions as an important molecular integrator that exerts its effects on developmental phase changes, at least in part through miRNA-regulated pathways [22].

The overexpression of *AtMIR156* causes the stimulation of PD, but *AtMIR172* diminishes it [13]. Interestingly, the expression of genes involved in miR156 and miR172 processing was lowered in *A. thaliana* and *L. sativa* dry viable seeds from *dog1* mutants, whereas the overexpression of MIR172 reduced seed PD [13]. It is highlighted that the essential role played by miR156 and miR172 in sensing and integrating environmental changes, and the role of DOG1 in affecting the relationship miR156/miR172 thus play a significant role in seed development [13]. Besides, small RNA library of Arabidopsis seeds obtained from the field in mid-summer (low dormancy) and mid-winter (high dormancy) possessed abundant miR156 levels at both seasonal periods. These data suggest that DOG1 maintains high miR156 levels in soil seed bank over the spatial sensing phase till PD release [33,34].

These and other recent revised results seem to fall under the hypothesis that DOG1 is part of a thermal sensing mechanism in the regulation of DOG1 transcription [1,34]. That is, DOG1 transduces environmental effects during maturation to alter depth of dormancy thus linking DOG1 and dormancy cycling. However, since *DOG1* expression follows environmental cues, DOG1 does not appear to directly determine the pattern of dormancy cycling [34,36]. In other words, although a key role of *DOG1* was detected in determining the depth of dormancy, it was still not a direct role for *DOG1* in generating altered seasonal patterns of seedling emergence [36]. Therefore, it remains unclear whether variation in DOG1 expression can drive variation in dormancy cycling behavior.

DOG1 has been recently shown to physically interact with two phosphatases (ABA-HYPERSENSITIVE GERMINATION 1 and 3; AHG1 and AHG3) to functionally block their essential downstream roles in the release of seed PD [25,42] (Figure 3). AHG1 and AHG3 belong to the clade-A type 2C PP2C family (i.e., with nine members in Arabidopsis) and act as negative regulators of ABA signaling and seed PD [25,42]. As a result, seed PD and ABA sensitivity are increased not only when DOG1 protein levels are enhanced [10] but also in an *ahg1 ahg3* double mutant [25]. Likewise, *ahg1-1ahg3-1hai3-1* triple mutant has a deeper dormancy and also proposed that at least AHG1, AHG3, and HAI3 are involved in the regulation of PD [42]. In addition, DOG1 controls PD by suppressing the action of specific PP2C phosphatases, which function as a convergence point of the ABA and DOG1 pathways [25]. On the other hand, the binding of DOG1 to heme group is essential for DOG1 task during PD [42]. Heme is an iron-binding protoporphyrin IX that regulates diverse biological activities (e.g., signal transduction); but the study of its role in ABA signaling is still in very early stages. Binding of DOG1 to AHG1 and heme seem to be an independent processes, though both are essential for DOG1 function in vivo [40]. Likewise, the role of the redox state of the cell in the control of seed PD is an unexplored research field [43]. We will return to these last topics later.

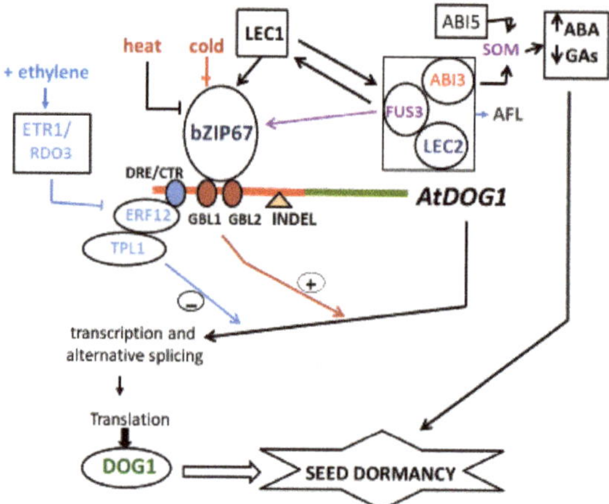

Figure 3. A model for the transcriptional regulation of *DOG1* gene by ethylene and low temperatures (promotor in red; coding region in green) to induce PD in *A. thaliana* (adapted from Li et al. 2019 [44]; Carbonero et al. 2017 [45], and Breeze et al. 2019 [46]). Downstream of the GBL1 *cis*-element is INDEL, a 285 bp sequence. INDEL is present in *DOG1* promoter of Ler-0 accession (weakly dormant). However, INDEL is absent in Cvi-0 ecotype (strong dormant). INDEL was found to directly affect the ability of bZIP67 to transactivate *DOG1* in vivo and may explain the observed variation in *DOG1* transcript levels, and consequently dormancy, between ecotypes. Such natural allelic variation in *DOG1* coupled with *DOG1* expression plasticity confers substantial adaptive significance in the field, where seasonal environmental factors can vary greatly, and the optimal timing of seed germination is primordial. ETR1: Ethylene Response-1; DRE/CRT: Dehydration-Responsive Element/C-Repeat; TPL1: Topless-1; ERF12: Ethylene Responsive TF-12; GBL1 and GBL2: G-Box-Like-1 and 2; SOM: Somnus.

2.3. Strict Regulation of DOG1 during Seed Dormancy

In *A. thaliana*, the *DOG1* gene is composed of 3 exons and 2 introns and is alternatively spliced of the second intron thus producing five transcript variants [26] and resulting in only three different proteins since translation of β and γ transcripts generate the same protein (Figure 2). The abundance of each *DOG1-mRNA* is different; but the proportion among them remains almost unaltered in different dormant and non-dormant accessions [2]. Recently, a short *DOG1-mRNA* was described [3]. This major form, named ε, excludes the third exon and incorporates a part of the second. The nascent mRNA-ε is formed by alternative polyadenylation and not by differential splicing [3,47]. The level of *DOG1* transcripts is strongly decreased throughout the final part of seed maturation, though the abundance of DOG1 protein does not diminish [10]. DOG1 α, β, γ, δ, and ε isoform proteins are functional [26]. DOG1-α, DOG1-β, and DOG1-δ were located in the nucleus and DOG1-β is much more abundant than the other isoforms [26]. Interestingly, the regulation of DOG1 protein accumulation by alternative splicing could be part of a mechanism to fine-tune seed dormancy (Figure 2). The DOG1 abundance might be explained by the existence of an altered ratio between *DOG1* transcripts. Thus, the abundance of the *DOG1-δ* at the final of maturation is higher compared to *DOG1-β*, which could increase the DOG1 protein although a general decrease in *DOG1* transcript levels is produced [26]. Interestingly, the solid results obtained by the Swiezewski's group suggest that the accumulation of short isoform of DOG1 ε-protein is enough to generate a dormancy phenotype. That is, the short *DOG1-mRNA* is translated in vivo and the generated DOG1 protein is also functional in controlling seed PD establishment [3]. The short ε-mRNA is the most abundant in *A. thaliana*, resulting in the same protein as β and γ-mRNA. This short isoform of DOG1 protein is better conserved than the other two, as the third exon is highly variable or even lost in other species and is also the most active in PD induction. On the other hand, Nakabayashi et al. [26] described that all the DOG1 isoforms show self-binding properties, and the loss of this capacity does not alter protein levels but leads to non-dormant phenotypes. In other words, DOG1 protein function is enhanced by its self-binding property. Thus, formation of dimers among different DOG1 isoforms would be necessary for the proper regulation of DOG1 functioning. Interestingly, *DOGL3* and *DOGL5* (*DOGL5.2*, alternative spliced form like *DOG1*) also bind to PP2C, and overexpression of *DOGL3* causes ABA hypersensitivity in seed germination, like *DOG1*, whereas *DOGL5* overexpression does not result in increased ABA sensitivity in seeds [40]. These and other features lead us to conclude that, *DOG1* and *DOGLs* do not seem to share a great deal of redundancy in terms of PD imposition.

It was shown that natural variation for DOG1 binding efficiency provides variation in PD [10,16,26]. These natural variations in several Arabidopsis accessions may be part of a system to adjust the PD [26]. However, more research is still required to clarify the exact role of the different DOG1 protein isoforms in plants. Finally, and given that *DOG1* is extensively regulated [1,3,9,13,25,42,47], recent relevant evidence demonstrated that asDOG1, a long non-coding antisense RNA from DOG1 in Arabidopsis, suppresses *DOG1* expression during seed maturation in *cis* and promotes germination [16,23]. This *DOG1* antisense partner originates close to the *DOG1* proximal polyadenylation. The presence of a conserved region in the second intron of *AtDOG1* led to the discovery of a natural noncoding antisense RNA (asRNA) (Figure 2). This conserved site, unexpected in a non-coding region, contains a promoter that extends approximately from the ending of the second exon to the transcription start site of *AtDOG1*. The transcription of *asDOG1* is independent of the *DOG1* promoter, and as it has been described for other genes, the asRNA acts as a negative regulator of *DOG1* sense transcription and expression [16]. Although the *asRNA* has a 5′ cap and is polyadenylated and stable, it seems, as said before, to regulate *DOG1* expression in *cis*, the antisense transcription process being more important as a repressor than the *asRNA* transcript itself [16]. The fact that *asDOG1* originates from close to the transcription termination site of the sense gene, raises the question of how this proximity affects antisense promoter activity. Although *asDOG1* is present in seeds, especially in desiccation phase, it is more abundantly transcribed in seedlings, meristems, and young leaves. The addition of ABA downregulates the transcription of *asDOG1* in vegetative tissues and allows *DOG1* expression. Moreover, when the promoter of *asDOG1* is mutated, the addition of ABA is unable to induce *DOG1* expression. This ABA-dependent *asDOG1* transcription is particularly important for the role of DOG1 in the induction of drought tolerance. Yatusevich et al. (2017) [23], in a sophisticated work on epigenetic regulation of DOG1 in *A. thaliana*, showed that, (i) *asDOG1* suppresses *DOG1* expression in both seeds (i.e., prevents PD) and leaves (i.e., drought response) and that *dog1* mutants are more susceptible to drought; (ii) *asDOG1* expression is powerfully inhibited by ABA; (iii) the ability of the antisense promotor to respond to ABA is absolutely required for the regulation of *DOG1* expression by ABA; (iv) this group advances the possibility that DOG1 may regulate PD through similar mechanisms involving the ABA pathway; and (v) in the absence of *asDOG1*, *DOG1* is constitutively highly expressed in leaves [23]. Among other deductions, it is interesting to conclude that DOG1, in addition to participating in seed PD, also has other roles in plants. Finally, *asDOG1*-mediated control of *DOG1* expression and seed PD appear to be *cis*-restricted, suggesting a mechanism that may involve *asDOG1* transcription rather than arising RNA [16,23].

3. ABA and Ethylene Signaling during DOG1 Tasks

3.1. ABA and DOG1 Work Cooperatively

The critical role of ABA in the induction of PD is currently unquestionable [33]. ABA and DOG1 seem to act in concert to carry out the PD, a striking event during seed development [10,47,48]. The action of DOG1 in regulating PD is ABA-coordinated [10]. That is, the regulation of PD by DOG1 requires a functional ABA signaling pathway [1]. Thus, although *dog1* dormancy phenotypes are analogous to ABA synthesis and signaling mutants (i.e., ABA sensitivity of *dog1* seeds is unchanged [10,12,42]), prevailing evidences suggest that DOG1 and ABA act in independent pathways. That is, the *dog1* mutant is completely non-dormant and lacks apparent pleiotropic phenotypes, indicating that DOG1 is a key player specific for the induction of PD. Interestingly, the integration of the near isogenic line NILDOG1-Cvi (i.e., strong *DOG1* expression) with the ABA-deficient mutant *aba1-1* (i.e., impaired in ABA biosynthesis and absence of seed dormancy), still results in the production of non-dormant seeds [9]. This finding suggests that ABA is indispensable for the DOG1 function in seed dormancy, although it is fair to remind that ABA cannot impose seed dormancy in the absence of DOG1 [9]. Together, ABA and DOG1 must be present to induce PD as absence of one of these two regulators results in complete lack of PD even when the other regulator is highly accumulated [9,12,18]. Likewise, different groups demonstrated that *AtDOG1* is also crucial for local adjustment to distinct environments [19]. Therefore, *DOG1* is unquestionably related to the natural alteration of PD in *A. thaliana*; but this has not yet been proven in other species [13]. Finally, the correlation between DOG1 and ABA metabolism, transport, and signaling are not properly known. This study deserves special attention.

3.2. ABI3, ABI5, and DOG1

ABI3 and *ABI5* encode essential ABA-dependent TFs and are two major components in the seed ABA signaling. DOG1 appears to control seed maturation and PD by controlling the expression of ABI5 (ABA INSENSITIVE 5; basic leucine zipper TF that functions in the core ABA signaling) and through genetic interaction with ABI3. These and other findings confirm that DOG1 is involved in the regulation of final phase of seeds development in coordination with ABA [9,13,23,25] (Figure 3). *ABI5* was not enough to suppress germination when overexpressed in Arabidopsis seeds and *abi5* shows a normal PD level [23]. Recent studies proposed that ABI5 is related to PD, and DOG1 regulates PD by controlling the *ABI5* expression [6]. Moreover, it was suggested that DOG1 stimulates ABI5 through the activation of seed maturation genes and the inhibition of germination-related transcripts [22]. Dekkers et al. (2016) [23] demonstrated that DOG1 interacts with PP2Cs, thus stimulating the expression of *ABI5* and PD. That is, the accumulation of ABA and ABI5 activity contributes to PD maintenance. In a nutshell, DOG1 regulates seed maturation and PD in part by controlling *ABI5* expression [23].

On the other hand, to determine the role of DOG1 in seed imbibition, several mutants related to the hormones synthesis and signaling have been analyzed, providing divergent results. For example, the ABA catabolic mutant *cyp77a2* exhibits high levels of ABA not only in dry seeds but also in seed imbibition, being highly dormant [49]. Nevertheless, the double mutant *dog1 cyp707a2* is less dormant than *cyp707a2* mutants, suggesting that the action of DOG1 is not completely ABA-dependent [10]. The Bentsink's group [50] described that the *dog1* mutants exhibit normal response to exogenous ABA, proposing that DOG1 is unable to control ABA signaling during seed imbibition. On the other hand, *dog1-5* mutant has very strong PD and increased expression of miRNA processing genes [22]. Studies with *dog1-5* mutant indicated that the short *DOG1*-mRNA is subject to alternative polyadenylation and that the short form of this transcript is functional. Likewise, this short DOG1 protein isoform is a key player to PD establishment [3]. It is interesting to note that in *abi5* mutants, the expression of *DOG1* was upregulated in the presence of ABA. This suggests a crosstalk between *DOG1* expression and ABA responses during seed imbibition [51]. Likewise, several authors have considered the positive effect of ABA on the *DOG1* expression during germination of many species [52]. In *L. sativum* seed germination, ABA increased *LesaDOG1*-mRNA, possibly involving ABI3 and ABI5 [11]. Similarly, in cruciferae *Sysimbrium officinale*, ABA shows a positive effect on the *SoDOG1* expression during early imbibition. Moreover, this positive effect may be inhibited by optimal AR whereas GAs can partly mimic the AR effect [7]. Experiments of chromatin immunoprecipitation (ChIP) showed no direct interaction of ABI3 with the promoter region of *DOG1* [53]. However, some relation seems to exist between *DOG1* and ABI3 as the double mutant *abi3-1 dog1-1* shows a severe *abi3* phenotype that is not present in the single *abi3-1* mutant [23]. This genetic interaction could be explained by ABI3 acting downstream of *DOG1*, rather than upstream, with the ABA probably involved. On the contrary, FUS3 ChIP-chip analysis revealed a direct interaction with *DOG1*, presumably involving the RY-element (CATGCATG) present in its promoter region [53,54] (Figure 3). Microarray experiments showed that FUS3 knockout mutants in Arabidopsis have a significant decrease in *DOG1* expression [55].

Within the puzzle that constitutes the functioning of DOG1, the possibility of acting as a TF is currently being evaluated [32]. But there is still no scientific evidence that DOG1 works as a TF. The possibility emerges from the analysis of the amino acid sequence of DOG1. Thus, the DOG1 family share sequence similarity with the TGACG motif-binding TFs (TGAs) and may have evolved from them, or vice versa. It is striking that all TGAs of bZIP TFs contain the DOG1 superfamily domain. However, TGA task in seed development is not clear. Quite recently, DOGL4 was isolated from seeds and some properties are already known [32]. For example, DOGL4 lacks a bZIP domain. In [32] it is suggested that DOGL4 (and possibly DOGL5) is the missing link between the DOG1 family proteins and TGAs [32]. Thus, (i) unlike DOG1, DOGL4 is induced by ABA; (ii) DOGL4 shares scarce homology in amino acid sequence with DOG1; (iii) DOGL4 plays a major role in mediating SSPs accumulation in seeds; that is, DOGL4 is an inducer of SSPs; (iv) *dog1* does not alter the majority of DOGL4-induced SSPs; and (v) seeds from *dogl4* mutant exhibit moderately enhanced PD. Current knowledge of DOGL4 and DOG1 suggests that at a time of evolution some of the properties of these two members of the DOG1 family may have diverged into two independent seed maturation regulators for distinct biochemical functions [32]. In other words, taken together with the knowledge about DOGL4, it can be suggested that DOG1 family proteins may have first evolved as seed maturation regulators. Finally, the induction of the seed maturation genes by DOG1 and DOGL4 may not be mediated through direct DNA binding, but probably represents indirect regulation because DOG1 is a heme-binding protein [25,32,42].

3.3. Ethylene and DOG1

The interaction of ET with other phytohormones and ROS in the seed life regulation is well documented [4,56]. Thus, in the same way as GAs and ABA, the crucial control of ET traffic in the cell and their coordination in specialized seed tissues is of considerable importance during PD and germination. Recently, Li et al. (2019) identified a module regulating PD in Arabidopsis [44]. This group demonstrated that DOG1 functions downstream of both ET receptor 1 (ETR1; RDO3) and ET-responsive transcription factor-12 (ERF12). Likewise, they also demonstrate that *RDO3* encodes *ETR1*, an ET receptor. In fact, ERF12 binds directly to the DOG1 promoter, recruiting the co-repressor

TOPLESS (TPL) in this nuclear process and inhibiting DOG1 expression. Likewise, through genetic analysis (i.e., *tpl-9 dog1-2* double mutant) this robust work insinuates that the regulation of PD by TPL depends on DOG1 [55]. In addition, (i) the transcriptional repressor in *A. thaliana* ERF12 (a member of the ERF-1B family) and TPL promote seed germination by repressing the DOG1 pathway; (ii) ERF12, functioning downstream of ETR1, is involved in regulating PD mediated by RDO3; (iii) ETR1 and ERF12 likely regulate PD through the DOG1 pathway [44]; and (iv) ERF12 binds to the promoter of *DOG1* and suppresses its expression (Figure 3). Together, perhaps DOG1 partially takes part in regulating PD mediated by the ET pathway. This possibility requires even more study. However, the role of the ETR1/RDO3-ERF12-TPL-DOG1 module discovered by the Li's group has clarified a notable part of the ET-controlled PD mechanism.

4. Transcription Factors Directly Involved in DOG1 Task

Although knowledge of DOG1 has advanced considerably in recent years, it is not known meticulously which TFs bind to the DOG1 promoter and are responsible for driving its expression during embryo maturation. Since the amount of DOG1 accumulated in the dry seed determines the storage time necessary to release PD, the regulation of *DOG1* transcriptional activity by ABI3, FUS3, LEC1, and LEC2 is an exploration that should be considered in detail. *DOG1* expression is known to rely indirectly on LEC1, a member of the HAP3 family [45,57,58]. However, LEC1 has no direct interaction with *DOG1* promoter (Figure 3). In *lec1* mutants, the gene activity of *DOG1* is reduced. ABI3, FUS3, and LEC2 (known as AFL as a whole) contain a B3 DNA-binding domain (special to plants) that specifically recognizes RY [CATGCA(TG)] motif present in the promoter region of many maturation-related genes [15,45,57]. A RY motif is present in the promoter of *AtDOG1*, and its transcriptional pattern in seed development indicates that *DOG1* is very likely regulated by at least one of above TFs [10]. In the attempt to identify some TF with affinity for the DOG1 promoter, it was perceived in soybean (*Glycine max*) that bZIP67 is a regulator of several genes involved in SSPs deposition [59]. In other words, *DOG1* encodes a plant specific protein with a domain shared by bZIP. Progressing in this approach, quite recently the Eastmond's group proved in a solid work that bZIP67 acts downstream LEC1 to transactivate DOG1 during seed maturation helping to establish PD in Arabidopsis [60]. Eastmond in his work demonstrated that: (i) bZIP67 is required for *DOG1* expression and DOG1 accumulation; (ii) probably, bZIP67 and DOG1 functionally belong to same pathway; (iii) bZIP67 may contribute to the regulation of PD through the control of *DOG1* expression; (iv) DOG1 is induced by LEC1; this fact occurs after the induction of bZIP67; (v) in vivo and in vitro experiments propose that bZIP67 binds to the promotor of DOG1 by GBL *cis*-element; (vi) cool conditions during seed maturation enhances bZIP67 amount but not bZIP67-mRNA [60]. Together, bZIP67 is a direct regulator of DOG1 expression, specifying the LEC1 action in the establishment of PD (Figure 3).

5. The Relationship between DOG1 and Protein Phosphatases

The major advances to unravel the molecular function of DOG1 are probably those provided by recent studies of direct interaction with other proteins. Pull-down experiments in vivo carried out by Née et al. [25] revealed 184 groups of proteins with direct interaction with DOG1. These results confirm again that DOG1 interacts with proteins involved in ABA responses [20,25]. Among these pulled down proteins there are some members of PP2C family [25,42]. PP2C phosphatases act as negative regulators of ABA signaling by inactivating sucrose nonfermenting-I-related protein kinases-2 (SnRK2), that act as positive effectors of ABA response. ABA receptors PYrabactin resistance-1 (PYR1)/PYR1-Like (PYL)/regulatory component of ABA receptor (RCAR) inactivate PP2Cs in the presence of ABA, allowing SnRK2s to trigger ABA-associated responses [48]. Two subfamilies of the PP2Cs (group-A) have been described in Arabidopsis: ABI1 and ABA-hypersensitive germination-1 (AHG1) [61]. The ABA receptors PYR1 interact with ABI1 subfamily members in the presence of ABA but not with AHG1 subfamily members, except for AHG3. On the contrary, DOG1 interacts with the AHG1 subfamily members but not with the ABI1 subfamily members [42]. These data suggest that PP2Cs could be the connection point between ABA and DOG1 signaling pathways. In addition to physically interacting, it has been demonstrated that *AHG1* and *AHG3* (i.e., both expressed in seeds with AHG1 being more specific to seeds than AHG3) are necessary for DOG1 functioning. Thus, while double mutant (*dog1-ahg1* and *dog1-ahg3*) show the same non-dormant phenotype than *dog1*, triple mutant (*dog1-ahg1-ahg3*) show highly dormant phenotype [25]. These data suggest that AHG1 and AHG3 act redundantly as negative regulators of PD induction by DOG1. The relationship between DOG1 and AHG1 has been studied in detail. Thus, it seems clear that DOG1 functions upstream of AHG1 in the ABA signaling pathway, and directly regulates the PP2C activity of AHG1 in an ABA-dependent manner [25]. Unlike AHG3, AHG1 does not interact with ABA receptors. In Arabidopsis, the deletion of residues 1-18 in the N-terminal region of DOG1 protein has a strong negative effect on both the interaction with AHG1 and the induction of PD [42]. This deletion prevents the ability of DOG1 to confer an ABA-hypersensitive phenotype, indicating that the interaction with AHG1 is indispensable for DOG1 function in both PD and germination control. Curiously, this small N-terminal region is not especially well conserved among species and it is unknown whether its deletion may affect the DOG1 conformation or only its interaction with AGH1 [14]. On the contrary, the deletion of 257-291 residues in the C-terminal region does not affect the interaction of DOG1 with AHG1 [42]. Very recently, the Nishimura's group demonstrates that the DSYLEW N-terminal sequence of DOG1 (spanning position 13–18), is essential for interacting with AHG1 [42]. Taken together, DOG1 enhances ABA signaling through its binding to AHG1 and AHG3 [25].

The presence of DOG1 decreases the AHG1 activity on its target SnRK2s. As in the case of PYL/PYR/RCAR, this indicates that DOG1 positively regulates the SnRK2s activity by inhibiting the PP2C activity of AHG1 [42]. It is likely that DOG1 has an effect on AHG3, similarly to the observed in AHG1, but until now the activity assays were carried out using artificial substrates for PP2C and it would be necessary to analyze its phosphatase activity specifically on SnRK2s [25,42]. So far, the existing data suggest that, (i) ABI1 phosphatase subfamily is regulated by PYR/PYL/RCAR receptor in presence of ABA; (ii) AHG1 is regulated by DOG1; and (iii) AHG3 is regulated by both ABA and DOG1. These interactions could explain the partial overlap between ABA and DOG1 mechanisms of action. At the same time, they could answer to the dormancy phenotype observed in *dog1* and ABA biosynthesis mutants [10]. In mutants with low expression of *DOG1*, the presence of ABA may not regulate the AHG1 activity, which continues acting as negative regulator of PD. On the contrary, in mutants with low ABA content, members of PP2C subfamily ABI1 remain active and would negatively regulate PD levels [42]. In addition to dephosphorylating SnRK2s, AHG1 also interacts with other proteins [42]. One of them is ABI FIVE BINDING PROTEIN 2 (AFP2), that acts as a negative regulator of the ABA response and interacts with ABI5 promoting its degradation [62]. These interactions of AHG1 could be important in the regulatory role of DOG1 too.

DOG1 has also shown to interact with reduced dormancy 5 (RDO5), a promoter of Arabidopsis PD, which belongs to PP2C family; but it lacks phosphatase activity. Curiously, mutant *rdo5* shows a reduced PD level, contrary to what occurs in mutant *ahg1-ahg3* [27,63]. If this pseudo phosphatase is involved in DOG1 signaling, its function would probably be different from other PP2Cs. It was demonstrated that RDO5 influences the PD analogously to DOG1 and, moreover, it mainly appears to function independent of ABA [63]. *RDO5* and *DOG1* have some similarities: (i) Seed preferential expression; (ii) positive regulators of PD; and (iii) their mutants exclusively show dormancy and germination defects without pleiotropic phenotypes [64]. The molecular mechanisms used by these dormancy-specific genes for regulating PD and their relationship with ABA signaling is still unclear and need to be studied more carefully [63]. Additionally, DOG1 also interacts with PROTEIN PHOSPHATASE 2A SUBUNIT A2 (i.e., PP2AA/PDF1) that acts as a negative regulator of PD. PDF1 acts upstream of DOG1 probably dephosphorylating DOG1 during the seed imbibition [27,65]. Concluding, all recent genetic analysis of this section indicate that: (i) The interaction of DOG1 with AHG1, AHG3, and RDO5 [25,42] was observed in the nucleus, whereas that the interaction with PDF1 (a DOG1 interacting phosphatase) occurs in the cytosol [25]; (ii) genetic analysis suggests that DOG1 acts as a suppressor of AHG1 and

AHG3 action in PD release; (iii) AHG1 and AHG3 have redundant roles in seed dormancy because the phenotype of the double mutant is much more severe than that of the single mutants; (iv) similar to *dog1-2*, the *pdf1* and *dog1-2* double mutant completely lacked PD. Therefore, *dog1-2* appears to be epistatic to *pdf1*, suggesting that DOG1 functions downstream of PDF1; and (v) ABA and DOG1 pathways converge at the level of the clade-A PP2C [20]. Interestingly, the loss of RDO5 function could be compensated for by low temperatures [20].

6. DOG1 also Undergoes Epigenetic Regulation

Chromatin modifications (e.g., methylation) are involved in the PD regulation; but little is known about its esential mechanism. It is interesting to highlight that methylation of histone 3 at Lys-4 (i.e., H3K4me3; active chromatin) in *DOG1* is more abundant in dormant seeds; while the repressing chromatin H3K27me3 predominates in germinating seeds. The polycomb repressing complex 1 (PRC1) seems to play a key role in these methylation changes. PRC1 interacts with the H3K4me3-binding ALFIN-like proteins and recruits PRC2 that establishes H3K27me3 marks [66]. In addition to the studies on *as*RNA and the reciprocal regulation of *DOG1-asDOG1* pair [16,67], the TFs HISTONE MONOUBIQUITYLATION-1 (HUB1), and REDUCED DORMANCY-2 (RDO2), involved in chromatin compaction, are required for the establishment of PD in *A. thaliana*. HUB1 and RDO2 are upregulated during seed maturation [68,69]. The *hub1* mutant have reduced seed dormancy [68]. Transcriptomic analysis of *hub1* (*rdo4*) and *rdo2* revealed a reduction of *DOG1* expression that could be associated to diminished PD [70]. Interestingly, PD level of *rdo2* increases whereas ABA content and sensitivity remain unchanged [4,63]. Moreover, *hub1* mutants obtained in a Landsberg *erecta* (Ler-0) accession were crossed with the near isogenic line NILDOG1 (i.e., strong DOG1 expression). The seeds from *hub1* mutants containing the DOG1-Cvi introgression were more dormant than those obtained from the original *hub1* mutant in a simple Ler background. The main conclusion of this experiment could be that *HUB1* is not epistatic to *DOG1* [68]. Therefore, *HUB1* is probably acting upstream of *DOG1* in combination with other regulatory mechanisms. In *rdo2* mutants, the addition of an extra copy of *DOG1* reversed the reduced PD back to wild type levels. The reduction of *DOG1* expression in this experiment is, at least in part, responsible for the dormancy phenotype observed in *rdo2* mutants [69,71]. Despite the above mentioned, *DOG1* is not necessarily regulated by HUB1 and RDO2 in a direct path. Many genes associated with maturation are also affected in *hub1* and *rdo2* mutants, including many genes associated with hormone metabolism and PD establishment, which can be mediating in the reduction of *DOG1* expression [69]. Contrary to *HUB1* and *RDO1*, the *KRYPTONITE/SUVH4* (*KYP*) gene encodes a methyltransferase involved in the regulation of PD acting as a negative regulator of the transcription processes [72,73]. *HUB* and *RDO1* are epistatic with respect to *KYP* [73]. Overexpression and mutations in *KYP* causes decreased and increased PD, respectively [73]. *KYP* gene is involved in the regulation of both ABA and GAs sensitivity in the seed and influences the transcriptional activity of DOG1 and other PD-related genes (e.g., *DOG1* and *ABI3*). Alternatively, KYP function could also be mediated by the alterations in the ABA: GAs balance rather than by an alteration of DOG1 histone methylation status. Interestingly, the analysis of double mutants *kyp-dog1*, *kyp-hub1*, and *kyp-rdo2* suggest that HUB1 and KYP act in the same pathway to regulate DOG1, while RDO2 acts in a parallel path [73]. In general, *HUB1* expression rises previously to the induction of PD, overlapping with an increase of *DOG1* transcription. It is striking that *KYP* expression is maximum in the summer when the PD and *DOG1* expression levels are low [32].

7. Are the Post-Translational Modifications of DOG1 Involved in Seed After-Ripening?

The PD can be relieved through a highly regulated process called AR that occurs in the dry viable seed [35,74,75]. AR upregulates the *sensu-stricto* germination modulating the sensitivity and metabolism of ABA [35,75]. The DOG1 protein becomes modified during AR, and its levels in stored viable seeds do not correlate with germination potential [10,12]. These protein modifications might prevent or reduce DOG1 self-binding and thereby its function [26]. The intensity of PD is directly related to DOG1 protein level accumulated in the dry mature viable seeds. This level defines the time that the seed must be stored to produce the PD release. DOG1 expression is associated with dormancy variation measured as AR time [9,26]. As a rule, the large amount of DOG1-mRNA accumulated in deep dormant seeds decreases during the onset of imbibition of both dormant and non-dormant seeds, but this decrease is faster in AR and germinating seeds [7,10]. Interestingly, the loss of PD by AR is not associated with an important reduction of the protein levels in the seed, but it could be related to a decrease of the DOG1 activity mediated by not yet known post-translational modifications [35,76]. Recently, it was proven that exogenous ABA increases the SoDOG1-mRNA levels. However, this increase is downregulated when optimal AR has been already established [7]. The inactivation of DOG1 might operate as a timer for overcoming PD and take part in a mechanism for controlling PD release [10,20]. Nakabayashi et al. (2012) proposed that post-translational protein modification (i.e., a shift in the isoelectric point; pI) makes DOG1 non-functional (i.e., inactivation) prior to and following AR [10]. These authors suggest a change in DOG1 structure and loss of function caused by AR. PDF1, a DOG1 interacting phosphatase, acts as a negative regulator of PD and its mutation prevents the isoelectrofocusing change of DOG1 during imbibition [10,25]. However, the germination of *pdf1* mutants is only slightly increased, and therefore other post-translational regulatory mechanisms may be necessary for the inactivation of DOG1. It is unknown how this process takes place, but it could be caused by a non-enzymatic oxidative process since, as mentioned above, they seem to have great importance in seed AR [43,75,77]. Consequently, the modifications experienced by DOG1 in the course of AR and the loss of PD might be due to oxidative processes [10]. In order to prove it, in a QTL mapping following the elevated partial pressure of oxygen (EPPO) treatment, *DOG1, DOG2,* and *DOG6* loci were identified. These loci had been previously identified employing AR to overcome PD [20,49]. As a conclusion to this approach it can be suggested that the release of PD by AR is principally produced by oxidative processes, and oxidative post-translational modifications of DOG1 could probably be associated with AR [20,65].

As indicated above, there is a negative correlation between the potential of germination and DOG1 protein levels [10,12]. But there are two important exclusions. On the one hand, the AR seeds may germinate even in high levels of DOG1 protein [10]. On the other hand, Nakabayashi et al. [26] showed that some seeds might germinate if they possess DOG1 unable to bind itself. In both cases, the absence of a negative correlation between the potential of germination and the accumulation of DOG1 protein can be due to similar causes. Accordingly, it is feasible that AR is related to the modifications of DOG1 protein which could affect its function, probably by avoiding or decreasing DOG1 self-binding [26]. Nevertheless, it is necessary to consider that DOG1 might act in PD signaling by an alternative pathway than in AR since it was proved that PD and AR seem to be independent at the molecular level [78]. Analysis of fully AR seeds in the Columbia (Col) background showed a decrease in ABA sensitivity for the *dog1-2* mutant [25].

8. Perspectives for Future Research on DOG1

Throughout this study we have reviewed a good number of recent publications related to DOG1, a protein absolutely required for the induction of PD. However, its molecular function is largely unknown. The hypothesis and conclusions that are currently underway in this review are largely based on data obtained from *A. thaliana* seed mutants and the characterization of cofactors that bind to different sites of the DOG1 structure. All known genetic and molecular data are part of a highly complex puzzle that, acting in a coordinated mode with ABA, trigger the generation of active DOG1 and appearance of PD. However, although the function of DOG1 seems conserved in Angiosperms, much more experimentation in monocot and dicot species needs to be done to confirm. The key to addressing the emergence and evolution of DOG1 involves studying similar genes in the DOG1 family (see Figure 1). Given the gaps existing in the knowledge of the mechanism of action of ABA in the PD process, this evolutionary study of DOG1 is of enormous importance since it firmly proves that DOG1 inhibits germination in an ABA-dependent way. However, ABA accumulates in seeds mainly to prevent viviparism and establish PD. The relationship between DOG1 and development of viviparism will be an important applied approach because reduced yield of certain crops is involved. On the other hand, the knowledge of endogenous and environmental signals responsible for the appearance/disappearance of DOG1 in the cells involved in the triggering, maintenance, and loss of PD, is an aspect still obscure and that requires a great deal of dedication. Recent research found that the production of miRNAs others than miR156 and miR172 (i.e., miR319, miR160, miR164, and miR390) are affected by DOG1 in lettuce and Arabidopsis. Although it seems clear that some miRNAs affect DOG1, it has not been conclusively proven yet that miRNA intervention is part of the primary mechanism by which DOG1 regulates PD. Hence, this intriguing and novel aspect should be taken into consideration to demonstrate the function of DOGs more comprehensively.

The yeast two-hybrid analysis (YTHA), co-immunoprecipitation, plant co-transformation, and epistatic analysis, are essential tools to progress in the knowledge of DOG1 at the molecular level. As a proof of the use of these methodologies, the breakthrough came from recent investigation about DOG1-interacting proteins (e.g., AHG1-DOG1). Although the interaction of DOG1 with AHG1 seems demonstrated, it is still not clear if the inactivation of AHG1 by DOG1 in vivo is a consequence of that interaction. Given that DOG1 affects the expression level of *ABI5*, the DOG1-AHG1 complex likely regulates ABI5 function. But this fact is still unsolved. Intriguingly, since the absorption spectrum of the active DOG1 exhibits characteristics of a heme-protein complex, it was proven that DOG1 binds to a heme group using His residues from its protein (Figure 4). It is striking that heme is not fundamental for DOG1 to interact with AHG1. Once it was known that heme group is involved in the action of DOG1, the PD mechanisms goes through a thorough investigation into the DOG1-heme relationship. Parallelly, at the evolutionary level, more research on ABA and heme signaling in the algal ancestor and bryophytes, are essential. However, the origin of heme in seeds is still an open question. Thus, is the plasmid only the source of heme in the seed cells or is there heme of mitochondrial origin?. This question opens an interesting line of research since the heme group has a remarkable interference in many aspects of seed physiology. Further studies are needed to elucidate the potential role of heme bound to DOG1 as a sensor of O_2 or NO. Thus, in the course of AR, several oxidative processes are produced, causing post-translational modifications in DOG1. It is necessary to investigate the possibility that ROS could oxidize DOG1. The nature of DOG1 modifications (e.g., phosphorylation/de-phosphorylation, redox changes, chromatin alterations, etc.,) in dry or imbibed AR viable seeds has not been demonstrated explicitly. These modifications may contribute to the change in the configuration of the DOG1 protein. Therefore, to carry out this approach, the study of the AR process in mutant seeds that are affected in the functioning of DOG1 can be an interesting tool. Finally, the knowledge in depth of PD physiology will imply to have tools for the control, among others, of crops with high agricultural value. In other words, the identification of DOG1 protein modifications is key to our understanding of PD and could enable the manipulation of PD levels in crop plants. Thus, a suitable understanding of PD will be advantageous for both ecological understanding and crop management. Likewise, it will be interesting to find out whether DOG1 is involved in the adaptation of PD to other environmental conditions that occur during seed maturation, like drought (Figure 5), light intensity, daylength, high or low temperatures, and nitrate levels. Together, the knowledge in depth of PD physiology will imply to have tools for the control, among others, of crops with high agricultural interest. Finally, to highlight the CRISPR/Cas (clustered regularly interspaced short palindromic repeats/CRISPR-associated protein) based genome editing approach has become a choice of technique due to its simplicity, ease of access, and flexibility [79,80]. The CRISPR/Cas9 system is a revolutionary technology for crop breeding and biological research through direct and controlled changes in the genome. Thus, the potential to edit multiple targets simultaneously makes CRIPSR/Cas9 possible to take up more challenging tasks required to engineer desired crop plants. The new gene editing techniques are more precise than standard genetic engineering tools that have been previously developed. The use of this technology will cooperate in the strong advance of knowledge of DOG1.

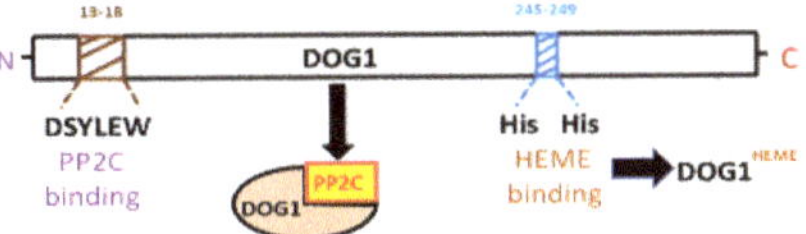

Figure 4. AtDOG1 protein structure indicating the positions of the AHG1 and heme binding sites. DOG1 is an α-helical protein that has the ability to bind both ABA HYPERSENSITIVE GERMINATION1 (AHG1; a clade A protein phosphatase 2C) and heme group. The amino acid residues that specifically bind PP2C (i.e., DSYLEW) are very close to the N-terminal; while those that specifically bind heme group are closer to the C-terminal. Heme binding is not necessary for DOG1 interaction with AHG1. However, heme binding at His245 and His249 is essential for DOG1 function in seed dormancy. Modified from Nonogaki (2019) [1,48].

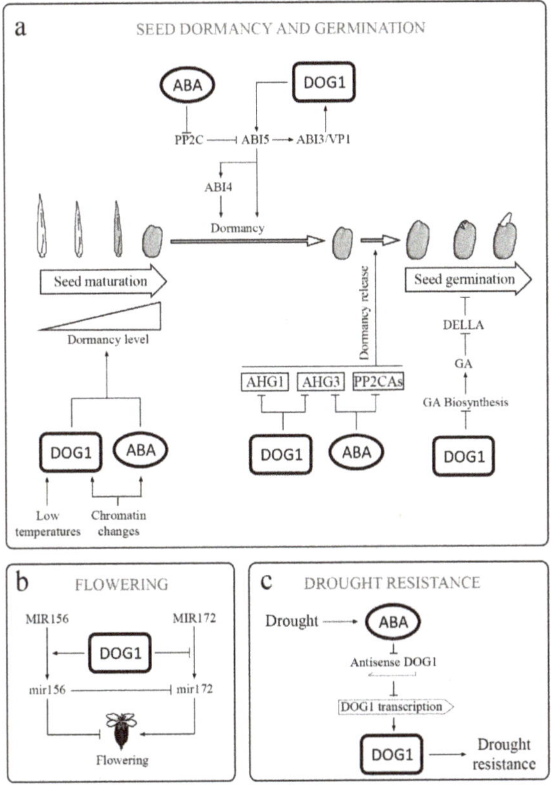

Figure 5. DOG1 in different physiological processes: (**a**) DOG1 has an influence on several genes involved in ABA signaling, including *ABI3* and ABI5. Thus, it acts together with ABA in the regulation of the acquisition, maintenance, and release of PD. DOG1 also inactivates AHG1 and AHG3, which are the key negative regulators in ABA signaling. The inactivation of AHG1 and AHG2 results in release of PD. Additionally, DOG1 reduces the GAs biosynthesis during germination. (**b**) DOG1 participates in the regulation of flowering by influencing the transition from primary MIR156 and MIR172 to active miR156 and miR172 through an effect on expression of genes involved in miRNA processing. Both active miRNAs are regulators, not only of flowering but also of PD and seed and seedling development. (**c**) Under drought conditions, the ABA content increases. The expression of DOG1 enhances the drought tolerance in vegetative tissues, and asDOG1 transcription is ABA repressed. asDOG1 acts as a negative regulator of DOG1 transcription.

Supplementary Materials: The following are available online at http://www.mdpi.com/2223-7747/9/4/480/s1, Figure S1: Alignment of deduced DOG1 and DOG1-like amino acid sequences used to create the phylogenetic tree (Figure 1). Sequences were aligned using MEGA-X software and the Clustal-W algorithm, Table S1: Genes and proteins included in the phylogenetic tree.

Author Contributions: N.C.-B., M.d.C.R.-G., and A.J.M. wrote the review. A.J.M. corrected and edited the final draft. All authors read and approved the final manuscript.

Funding: This research received no external funding.

Acknowledgments: We thank Tania García Lamas for critical reading and advice.

Conflicts of Interest: The authors declare no conflict of interest.

References

1. Nonogaki, H. Seed germination and dormancy: The classic story, new puzzles, and evolution. *J. Integr. Plant Biol.* **2019**, *61*, 541–563. [CrossRef] [PubMed]

2. Graeber, K.; Voegele, A.; Büttner-Mainik, A.; Sperber, K.; Mummenhoff, K.; Leubner-Metzger, G. Spatiotemporal seed development analysis provides insight into primary dormancy induction and evolution of the *Lepidium DELAY OF GERMINATION 1* genes. *Plant Physiol.* **2013**, *161*, 1903–1917. [CrossRef] [PubMed]

3. Cyrek, M.; Fedak, H.; Ciesielski, A.; Guo, Y.; Sliwa, A.; Brzezniak, L.; Krzyczmonik, K.; Pietras, Z.; Kaczanowski, S.; Liu, F.; et al. Seed dormancy in Arabidopsis is controlled by alternative polyadenylation of *DOG1*. *Plant Physiol.* **2016**, *170*, 947–955. [CrossRef] [PubMed]

4. Shu, K.; Liu, X.D.; Xie, Q.; He, Z.H. Two faces of one seed: Hormonal regulation of dormancy and germination. *Mol. Plant* **2016**, *9*, 34–45. [CrossRef]

5. Ashikawa, I.; Mori, M.; Nakamura, S.; Abe, F.A. A transgenic approach to controlling wheat seed dormancy level by using Triticeae *DOG1-like* genes. *Transgenic Res.* **2014**, *23*, 621–629. [CrossRef]

6. Zhao, M.; Yang, S.; Liu, X.; Wu, K. Arabidopsis histone demethylases LDL1 and LDL2 control primary seed dormancy by regulating *DELAY OF GERMINATION 1* and ABA signaling-related genes. *Front. Plant Sci.* **2015**, *6*, 159. [CrossRef]

7. Carrillo-Barral, N.; Matilla, A.J.; García-Ramas, C.; Rodríguez-Gacio, M.C. ABA-stimulated *SoDOG1* expression is after-ripening inhibited during early imbibition of germinating *Sisymbrium officinale* seeds. *Physiol. Plant.* **2015**, *155*, 457–471. [CrossRef]

8. Alonso-Blanco, C.; Bentsink, L.; Hanhart, C.J.; Blankestijn-de Vries, H.; Koornneef, M. Analysis of natural allelic variation at seed dormancy loci of *Arabidopsis thaliana*. *Genetics* **2003**, *164*, 711–729.

9. Bentsink, L.; Jowett, J.; Hanhart, C.J.; Koornneef, M. Cloning of DOG1, a quantitative trait locus controlling seed dormancy in Arabidopsis. *Proc. Natl. Acad. Sci. USA* **2006**, *103*, 17042–17047. [CrossRef]

10. Nakabayashi, K.; Bartsch, M.; Xiang, Y.; Miatton, E.; Pellengahr, S.; Yano, R.; Seo, M.; Soppe, W.J.J. The time required for dormancy release in *Arabidopsis* is determined by DELAY OF GERMINATION1 protein levels in freshly harvested seeds. *Plant Cell* **2012**, *24*, 2826–2838. [CrossRef]

11. Graeber, K.; Linkies, A.; Müller, K.; Wunchova, A.; Rott, A.; Leubner-Metzger, G. Cross-species approaches to seed dormancy and germination: Conservation and biodiversity of ABA-regulated mechanisms and the Brassicaceae *DOG1* genes. *Plant Mol. Biol.* **2010**, *73*, 67–87. [CrossRef]

12. Graeber, K.; Linkies, A.; Steinbrecher, T.; Mummenhoff, K.; Tarkowska, D.; Tureckova, V.; Ignatz, M.; Sperber, K.; Voegele, A.; de Jong, H.; et al. *DELAY OF GERMINATION 1* mediates a conserved coat-dormancy mechanism for the temperature-and gibberellin-dependent control of seed germination. *Proc. Natl. Acad. Sci. USA* **2014**, *111*, E3571–E3580. [CrossRef]

13. Huo, H.; Wei, S.; Bradford, K.J. DELAY OF GERMINATION1(DOG1) regulates both seed dormancy and flowering time through microRNA pathways. *Proc. Natl. Acad. Sci. USA* **2016**, *113*, E2199–E2206. [CrossRef]

14. Ashikawa, A.; Abe, F.; Nakamura, S. *DOG1-like* genes in cereals: Investigation of their function by means of ectopic expression in Arabidopsis. *Plant Sci.* **2013**, *208*, 1–9. [CrossRef]

15. Rikiishi, K.; Maekawa, M. Seed maturation regulators are related to the control of seed dormancy in wheat (*Triticum aestivum* L.). *PLoS ONE* **2014**, *9*, e107618. [CrossRef]

16. Fedak, H.; Palusinska, M.; Krzyczmonik, K.; Brzezniak, L.; Yatusevich, R.; Pietras, Z.; Kaczanowski, S.; Swiezewski, S. Control of seed dormancy in Arabidopsis by a cis-acting noncoding antisense transcript. *Proc. Natl. Acad. Sci. USA* **2016**, *113*, E7846–E7855. [CrossRef]

17. Chiang, G.C.K.; Bartsch, M.; Barua, D.; Nakabayashi, K.; Debieu, M.; Kronholm, I.; Koornneef, M.; Soppe, W.J.J.; Donohue, K.; De Meaux, J. *DOG1* expression is predicted by the seed-maturation environment and contributes to geographical variation in germination in Arabidopsis thaliana. *Mol. Ecol.* **2011**, *20*, 3336–3349.

18. Kendall, S.L.; Hellwege, A.; Marriot, P.; Whalley, C.; Graham, I.A.; Penfield, S. Induction of dormancy in *Arabidopsis* summer annuals requires parallel regulation of *DOG1* and hormone metabolism by low temperature and CBF transcription factors. *Plant Cell.* **2011**, *23*, 2568–2580. [CrossRef]

19. Murphey, M.; Kovach, K.; Elnaccash, T.; He, H.; Bentskink, L.; Donohue, K. DOG1-imposed dormancy mediates germination responses to temperature cues. *Environ. Exp. Bot.* **2015**, *112*, 33–43. [CrossRef]

20. Née, G.; Xiang, Y.; Soppe, W.J. The release of dormancy, a wake-up call for seeds to germinate. *Curr. Opin. Plant Biol.* **2017**, *35*, 8–14. [CrossRef]

21. Buijs, G.; Kodde, J.; Groot, S.P.C.; Bentsink, L. Seed dormancy release accelerated by elevated partial pressure of oxygen is associated with DOG loci. *J. Exp. Bot.* **2018**, *69*, 3601–3608. [CrossRef]

22. Yatusevich, R.; Fedak, H.; Ciesielski, A.; Krzyczmonik, K.; Kulik, A.; Dobrowolska, G.; Swiezewski, S. Antisense transcription represses *Arabidopsis* seed dormancy QTL *DOG1* to regulate drought tolerance. *EMBO Rep.* **2017**, *18*, e201744862. [CrossRef]

23. Dekkers, B.J.; He, H.; Hanson, J.; Willems, L.A.; Jamar, D.C.; Cueff, G.; Rajjou, L.; Hilhorst, H.W.; Bentsink, L. The Arabidopsis DELAY OF GERMINATION 1 gene affects ABSCISIC ACID INSENSITIVE 5 (ABI5) expression and genetically interacts with ABI3 during Arabidopsis seed development. *Plant J.* **2016**, *85*, 451–465. [CrossRef]

24. Auge, G.A.; Penfield, S.; Donohue, K. Pleiotropy in developmental regulation by flowering-pathway genes: Is it an evolutionary constraint? *New Phytol.* **2019**, *224*, 55–70. [CrossRef]

25. Née, G.; Kramer, K.; Nakabayashi, K.; Yuan, B.; Xiang, Y.; Miatton, E.; Finkemeier, I.; Soppe, W.J.J. DELAY OF GERMINATION1 requires PP2C phosphatases of the ABA signaling pathway to control seed dormancy. *Nat. Commun.* **2017**, *8*, 72. [CrossRef]

26. Nakabayashi, K.; Bartsch, M.; Ding, J.; Soppe, W.J.J. Seed dormancy in Arabidopsis requires self-binding ability of DOG1 protein and the presence of multiple isoforms generated by alternative splicing. *PLoS Genet.* **2015**, *11*, e1005737. [CrossRef]

27. Gutierrez, L.; Van Wuytswinkel, O.; Castelain, M.; Bellini, C. Combined networks regulating seed maturation. *Trends Plant Sci.* **2007**, *12*, 294–300. [CrossRef]

28. Bewley, J.D.; Nonogaki, H. Seed Maturation and Germination. In *Reference Module in Life Sciences*; Elsevier: North York, ON, Canada, 2017. [CrossRef]

29. Zinsmeister, J.; Lalanne, D.; Terrasson, E.; Chatelain, E.; Vandecasteele, C.; Vu, B.L.; Dubois-Laurent, C.; Geoffriau, E.; Le Signor, C.; Dalmais, M.; et al. ABI5 is a regulator of seed maturation and longevity in legumes. *Plant Cell* **2016**, *28*, 2735–2754. [CrossRef]

30. Leprince, O.; Pellizzaro, A.; Berriri, S.; Buitink, J. Late seed maturation: Drying without dying. *J. Exp. Bot.* **2017**, *68*, 827–841. [CrossRef]

31. Li, P.; Ni, H.; Ying, S.; Wei, J.; Hu, X. Teaching an old dog a new trick: multifaceted strategies to control primary seed germination by DELAY OF GERMINATION 1 (DOG1). *Phyton-Int. J. Exp. Bot.* **2020**, *89*, 1–12. [CrossRef]

32. Sall, K.; Dekkers, B.J.W.; Nonogaki, M.; Katsuragawa, Y.; Koyari, R.; Hendrix, D.; Willems, L.A.J.; Bentsink, L.; Nonogaki, H. DELAY OF GERMINATION 1-LIKE 4 acts as an inducer of seed reserve accumulation. *Plant J.* **2019**, *100*, 7–19. [CrossRef]

33. Seo, M.; Marion-Poll, A. Abscisic acid metaboolism and transport. *Adv. Bot. Res.* **2019**, *92*, 2–41.

34. Finch-Savage, W.E.; Footitt, S. Seed dormancy cycling and the regulation of dormancy mechanisms to time germination in variable field environments. *J. Exp. Bot.* **2017**, *68*, 843–856. [CrossRef]

35. Matilla, A.J.; Carrillo-Barral, N.; Rodríguez-Gacio, M.C. An Update on the Role of NCED and CYP707A ABA metabolism genes in seed dormancy induction and the response to after-ripening and nitrate. *J. Plant Growth Regul.* **2015**, *34*, 274–293. [CrossRef]

36. Footitt, S.; Walley, P.G.; Lynn, J.R.; Hambidge, A.J.; Penfield, S.; Finch-Savage, W.E. Trait analysis reveals DOG1 determines initial depth of seed dormancy, but not changes during dormancy cycling that result in seedling emergence timing. *New Phytol.* **2020**, *225*, 2035–2047. [CrossRef]

37. Postma, F.M.; Lundemo, S.; Ågren, J. Seed dormancy cycling and mortality differ between two locally adapted populations of *Arabidopsis thaliana*. *Ann. Bot.* **2016**, *117*, 249–256.

38. Postma, F.M.; Agren, J. Early life stages contribute strongly to local adaptation in *Arabidopsis thaliana*. *Proc. Natl. Acad. Sci. USA* **2016**, *113*, 7590–7595. [CrossRef]

39. Née, G.; Obeng-Hinneh, E.; Sarvari, P.; Nakabayashi, K.; Soppe, W.J.J. Secondary dormancy in *Brassica napus* is correlated with enhanced *BnaDOG1* transcript levels. *Seed Sci. Res.* **2015**, *25*, 221–229.

40. Bazin, J.; Langlade, N.; Vincourt, P.; Arribat, S.; Balzergue, S.; El-Maarouf-Bouteau, H.; Bailly, C. Targeted mRNA oxidation regulates sunflower seed dormancy alleviation during dry after-ripening. *Plant Cell* **2011**, *23*, 2196–2208. [CrossRef]

41. Huang, D.; Koh, C.; Feurtado, J.A.; Tsang, E.W.; Cutler, A.J. MicroRNAs and their putative targets in *Brassica napus* seed maturation. *BMC Genom.* **2013**, *14*, 140. [CrossRef]

42. Nishimura, N.; Tsuchiya, W.; Moresco, J.J.; Hayashi, Y.; Satoh, K.; Kaiwa, N.; Irisa, T.; Kinoshita, T.; Schroeder, J.I.; Yates, J.R.; et al. Control of seed dormancy and germination by DOG1-AHG1 PP2C phosphatase complex via binding to heme. *Nat. Commun.* **2018**, *9*, 2132. [CrossRef] [PubMed]

43. Bailly, C. The signalling role of ROS in the regulation of seed germination and dormancy. *Biochem. J.* **2019**, *476*, 3019–3032. [CrossRef] [PubMed]

44. Li, X.; Chen, T.; Li, Y.; Wang, Z.; Cao, H.; Chen, F.; Li, Y.; Soppe, W.J.J.; Li, W.; Liu, Y. ETR1/RDO3 regulates seed dormancy by relieving the inhibitory effect of the ERF12-TPL complex on DELAY OF GERMINATION1 expression. *Plant Cell* **2019**, *31*, 832–884. [CrossRef] [PubMed]

45. Carbonero, P.; Iglesias-Fernández, R.; Vicente-Carbajosa, J. The AFL subfamily of B3 transcription factors: Evolution and function in angiosperm seeds. *J. Exp. Bot.* **2017**, *68*, 871–880. [CrossRef]

46. Breeze, E. Letting sleeping DOGs lie: Regulation of DOG1 during seed dormancy. *Plant Cell.* **2019**, *31*, 1218–1219. [CrossRef]

47. Szakonyi, D.; Duque, P. Alternative splicing as a regulator of early plant development. *Front. Plant Sci.* **2018**, *19*, 1174. [CrossRef]

48. Nonogaki, H. ABA responses during seed development and germination. *Adv. Bot. Res.* **2019**, *92*. [CrossRef]

49. Okamoto, M.; Kuwahara, A.; Seo, M.; Kushiro, T.; Asami, T.; Hirai, N.; Kamiya, Y.; Koshiba, T.; Nambara, E. CYP707A1 and CYP707A2, which encode abscisic acid 8′-hydroxylases, are indispensable for proper control of seed dormancy and germination in Arabidopsis. *Plant Physiol.* **2006**, *141*, 97–107. [CrossRef]

50. Bentsink, L.; Hanson, J.; Hanhart, C.J.; de Vries, H.B.; Coltrane, C.; Keizer, P.; El-Lithy, M.; Alonso-Blanco, C.; de Andres, M.T.; Reymond, M. Natural variation for seed dormancy in Arabidopsis is regulated by additive genetic and molecular pathways. *Proc. Natl. Acad. Sci. USA* **2010**, *107*, 4264–4269. [CrossRef]

51. Kinoshita, N.; Berr, A.; Belin, C.; Chappuis, R.; Nishizawa, N.K.; Lopez-Molina, L. Identification of growth insensitive to *ABA3* (*gia3*), a recessive mutation affecting ABA signaling for the control of early postgermination growth in *Arabidopsis thaliana*. *Plant Cell Physiol.* **2010**, *51*, 239–251. [CrossRef]

52. Teng, S.; Rognoni, S.; Bentsink, L.; Smeekens, S. The Arabidopsis GSQ5/DOG1 Cvi allele is induced by the ABA-mediated sugar signalling pathway and enhances sugar sensitivity by stimulating ABI4 expression. *Plant J.* **2008**, *55*, 372–381. [CrossRef] [PubMed]

53. Mönke, G.; Altschmied, L.; Tewes, A.; Reidt, W.; Mock, H.P.; Bäumlein, H.; Conrad, U. Seed-specific transcription factors ABI3 and FUS3: Molecular interaction with DNA. *Planta* **2004**, *219*, 158–166. [CrossRef] [PubMed]

54. Wang, F.; Perry, S.E. Identification of direct targets of FUSCA3, a key regulator of *Arabidopsis thaliana* seed development. *Plant Physiol.* **2013**, *161*, 1251–1264. [CrossRef] [PubMed]

55. Yamamoto, A.; Kagaya, Y.; Usui, H.; Hobo, T.; Takeda, S.; Hattori, T. Diverse roles and mechanisms of gene regulation by the Arabidopsis seed maturation master regulator FUS3 revealed by microarray analysis. *Plant Cell Physiol.* **2010**, *51*, 2031–2046. [CrossRef]

56. Nonogaki, H. Seed biology updates—Highlights and new discoveries in seed dormancy and germination research. *Front Plant Sci.* **2017**, *8*, 524. [CrossRef]

57. Giraudat, J.; Hauge, B.M.; Valon, C.; Smalle, J.; Parcy, F.; Goodman, H.M. Isolation of the Arabidopsis *ABI3* gene by positional cloning. *Plant Cell* **1992**, *4*, 1251–1261.

58. Pelletier, J.M.; Kwong, R.W.; Park, S.; Le, B.H.; Baden, R.; Cagliari, A.; Hashimoto, M.; Muñoz, M.D.; Fischer, F.L.; Goldberg, R.B.; et al. LEC1 sequentially regulates the transcription of genes involved in diverse developmental processes during seed development. *Proc. Natl. Acad. Sci. USA* **2017**, *114*, E6710–E6719. [CrossRef]

59. Dröge-Laser, W.; Snoek, B.L.; Snel, B.; Weiste, C. The Arabidopsis bZIP transcription factor family—An update. *Curr. Opin. Plant Biol.* **2018**, *45*, 36–49. [CrossRef]

60. Bryant, F.N.; Hudges, D.; Hassani-Pak, K.; Eastmond, P.J. Basic LEUCINE ZIPPER TRANSCRIPTION FACTOR67 transactivates DELAY OF GERMINATION1 to establish primary seed dormancy in Arabidopsis. *Plant Cell* **2019**, *31*, 1276–1288. [CrossRef]

61. Xue, T.; Wang, D.; Zhang, S.; Ehlting, J.; Ni, F.; Jakab, S.; Zeng, C.; Zhong, Y. Genome-wide and expression analysis of protein phosphatase 2C in rice and Arabidopsis. *BMC Genomics* **2008**, *9*, 550. [CrossRef]

62. Garcia, M.E.; Lynch, T.; Peeters, J.; Snowden, C.; Finkelstein, R. A small plant specific protein family of ABI five binding proteins (AFPs) regulates stress response in germinating Arabidopsis seeds and seedlings. *Plant Mol. Biol.* **2008**, *67*, 643–658. [CrossRef] [PubMed]

63. Xiang, Y.; Nakabayashi, K.; Ding, J.; He, F.; Bentsink, L.; Soppe, W.J.J. Reduced dormancy 5 encodes a protein phosphatase 2C that is required for seed dormancy in *Arabidopsis*. *Plant Cell* **2014**, *26*, 4362–4375. [CrossRef] [PubMed]

64. Kerdaffrec, E.; Filiault, D.L.; Korte, A.; Sasaki, E.; Nizhynska, V.; Seren, Ü.; Nordborg, M. Multiple alleles at a single locus control seed dormancy in Swedish Arabidopsis. *eLife* **2016**, *5*, e22502. [CrossRef] [PubMed]

65. Zhou, H.W.; Nussbaumer, C.; Chao, Y.; DeLong, A. Disparate roles for the regulatory A subunit isoforms in Arabidopsis protein phosphatase 2A. *Plant Cell* **2004**, *16*, 709–722. [CrossRef] [PubMed]

66. Molitor, A.M.; Bu, Z.; Yu, Y.; Shen, W.H. Arabidopsis AL PHD-PRC1 complexes promote seed germination through H3K4me3-to-H3K27me3 chromatin state switch in repression of seed developmental genes. *PLoS Genet.* **2014**, *10*, e1004091. [CrossRef] [PubMed]

67. Kowalczyk, J.; Palusinska, M.; Wroblewska-Swiniarska, A.; Pietras, Z.; Szewc, L.; Dolata, J.; Jarmolowski, A.; Swiezewski, S. Alternative polyadenylation of the sense transcript controls antisense transcription of Arabidopsis seed dormancy QTL gene DOG1. *Mol. Plant* **2017**, *10*, 1349–1352. [CrossRef]

68. Liu, Y.; Koornneef, M.; Soppe, W.J.J. The absence of histone H2B monoubiquitination in the *Arabidopsis hub1 (rdo4)* mutant reveals a role for chromatin remodeling in seed dormancy. *Plant Cell* **2007**, *19*, 433–444. [CrossRef]

69. Liu, Y.; Geyer, R.; van Zanten, M.; Carles, A.; Li, Y.; Hörold, A.; van Nocker, S.; Soppe, W.J.J. Identification of the Arabidopsis REDUCED DORMANCY 2 gene uncovers a role for the polymerase-associated factor 1 complex in seed dormancy. *PLoS ONE* **2011**, *6*, e22241. [CrossRef]

70. Mortensen, S.A.; Sønderkær, M.; Lynggaard, C.; Grasser, M.; Nielsen, K.L.; Grasser, K.D. Reduced expression of the DOG1 gene in Arabidopsis mutant seeds lacking the transcript elongation factor TFIIS. *FEBS Lett.* **2011**, *585*, 1929–1933. [CrossRef]

71. Mortensen, S.A.; Grasser, K.D. The seed dormancy defect of Arabidopsis mutants lacking the transcript elongation factor TFIIS is caused by reduced expression of the DOG1 gene. *FEBS Lett.* **2014**, *588*, 47–51. [CrossRef]

72. Jackson, J.P.; Johnson, L.; Jasencakova, Z.; Zhang, X.; Perez-Burgos, L.; Singh, P.B.; Cheng, X.; Schubert, I.; Jenuwein, T.; Jacobsen, S.E. Dimethylation of histone H3 lysine 9 is a critical mark for DNA methylation and gene silencing in *Arabidopsis thaliana*. *Chromosoma* **2004**, *112*, 308–315. [CrossRef] [PubMed]

73. Zheng, J.; Chen, F.; Wang, Z.; Cao, H.; Li, X.; Deng, X.; Soppe, W.J.J.; Li, Y.; Liu, Y. A novel role for histone methyltransferase KYP/SUVH4 in the control of Arabidopsis primary seed dormancy. *New Phytol.* **2012**, *193*, 605–616. [CrossRef] [PubMed]

74. Dekkers, B.J.; Pearce, S.P.; van Bolderen-Veldkamp, R.P.M.; Holdsworth, M.J.; Bentsink, L. Dormant, after-ripened *Arabidopsis thaliana* seeds are distinguished by early transcriptional differences in the imbibed state. *Front. Plant Sci.* **2016**, *7*, 1323–1338. [CrossRef] [PubMed]

75. Iglesias-Fernández, R.; Rodríguez-Gacio, M.C.; Matilla, A.J. Progress in research on dry afterripening. *Seed Sci. Res.* **2011**, *21*, 69–80. [CrossRef]

76. Chahtane, H.; Kim, W.; Lopez-Molina, L. Primary seed dormancy: A temporally multilayered riddle waiting to be unlocked. *J. Exp. Bot.* **2017**, *68*, 857–869. [CrossRef]

77. Morscher, F.; Kranner, I.; Arc, E.; Bailly, C.; Roach, T. Glutathione redox state, tocochromanols, fatty acids, antioxidant enzymes and protein carbonylation in sunflower seed embryos associated with after-ripening and ageing. *Ann. Bot.* **2015**, *116*, 669–678. [CrossRef]
78. Carrera, E.; Holman, T.; Medhurst, A.; Dietrich, D.; Footitt, S.; Theodoulou, F.L.; Holdsworth, M.J. Seed after-ripening is a discrete developmental pathway associated with specific gene networks in Arabidopsis. *Plant J.* **2008**, *53*, 214–224. [CrossRef]
79. Vats, S.; Kumawat, S.; Kumar, V.; Patil, G.B.; Joshi, T.; Sonah, H.; Sharma, T.R.; Deshmukh, R. Genome editing in plants: Exploration of technological advancements and challenges. *Cells* **2019**, *8*, 1386. [CrossRef]
80. Corte, L.E.D.; Mhmaoud, L.M.; Moraes, T.S.; Mou, Z.; Grosser, J.W.; Dutt, M. Development of improved fruit, vegetable, and ornamental crops using the CRISPR/Cas9 genome editing technique. *Plants* **2019**, *8*, 601. [CrossRef]

MDPI

St. Alban-Anlage 66

4052 Basel

Switzerland

Tel. +41 61 683 77 34

Fax +41 61 302 89 18

www.mdpi.com

Plants Editorial Office

E-mail: plants@mdpi.com

www.mdpi.com/journal/plants